W9-CNF-518

Lecture Notes in Mathematics

Edited by A. Dold and B. Eckmann

464

Charles Rockland

Hypoellipticity and Eigenvalue Asymptotics

Springer-Verlag
Berlin · Heidelberg · New York 1975

Author
Prof. Charles Rockland
Brandeis University
Department of Mathematics
Waltham, Massachusetts, 02154
USA

Library of Congress Cataloging in Publication Data

Rockland, C 1947-
Hypoellipticity and eigenvalue asymptotics.

(Lecture notes in mathematics ; 464)
Includes bibliographical references and index.
1. Differential equations, Partial. 2. Differential equations, Hypoelliptic. 3. Eigenvalues. 4. Asymptotic expansions. I. Title. II. Series: Lecture notes in mathematics (Berlin) ; 464.
QA3.L28 no.464 [QA377] 510'.8s [515'.353]
75-16382

AMS Subject Classifications (1970): 22E25, 22E45, 35D05, 35D10, 35H05, 35N15, 35P15, 58G15

ISBN 3-540-07175-X Springer-Verlag Berlin · Heidelberg · New York
ISBN 0-387-07175-X Springer-Verlag New York · Heidelberg · Berlin

Printed in Germany
Offsetdruck: Julius Beltz, Hemsbach/Bergstr.

Table of Contents

§1. Introduction

In this paper we examine local solvability and hypo-ellipticity properties of certain pseudo-differential operators P with multiple characteristics from the standpoint of the test-operators. These test-operators are geometric invariants of P which may be regarded as first-order approximations to some type of intrinsic partial Fourier transform of P. Alternately, they may constitute an appropriate notion of principal symbol for P. To begin our discussion, we shall give a rather sketchy survey of some of the recent work in this area. This will be followed by a description of the results to be found in the subsequent sections of the paper.

Grushin in [13] and [14] studies certain differential and pseudo-differential operators $P(\underline{y},D_{\underline{y}},D_{\underline{x}})$ of special type, satisfying certain quasi-homogeneity conditions. He shows that the local solvability and hypo-ellipticity of the operators $P(\underline{y},D_{\underline{y}},D_{\underline{x}})$ is related to eigenvalue properties of the operators $P(\underline{y},D_{\underline{y}},\underline{\xi})$ obtained by partial Fourier transformation of

$P(\underline{y},D_{\underline{y}},D_{\underline{x}})$ in the $\underline{x}$-variables. In particular he proves that $P(\underline{y},D_{\underline{y}},D_{\underline{x}})$ is hypoelliptic if and only if for every $\underline{\xi}$ such that $|\underline{\xi}| = 1$, $\underline{P}(\underline{y},D_{\underline{y}},\underline{\xi})$ does not have 0 as an eigenvalue. Grushin also shows that the difference between local solvability and hypoellipticity for the operator $P(\underline{y},D_{\underline{y}},D_{\underline{x}})$ is measured by the index of the partial Fourier transformed operators $P(\underline{y},D_{\underline{y}},\xi)$, $|\underline{\xi}| = 1$. Moreover, under various boundary and coboundary conditions related to the above index, he shows how to construct right and left parametrices for $P(\underline{y},D_{\underline{y}},D_{\underline{x}})$.

Grushin, although he derives the eigenvalue criterion does not explicitly determine the eigenvalues of the operator $\underline{P}(\underline{y},D_{\underline{y}},\underline{\xi})$ except in some special cases. Gilioli and Treves in [12], using methods quite different from those of Grushin, derive explicit conditions that are necessary and sufficient for local solvability at the origin in R^2 of the operator.

$$P = \left(\frac{\partial}{\partial t} - iat^k \frac{\partial}{\partial x}\right)\left(\frac{\partial}{\partial t} - ibt^k \frac{\partial}{\partial x}\right) + ict^{k-1} \frac{\partial}{\partial x} \tag{1.1}$$

Here k is an odd integer and a,b,c are assumed real.

In particular, in the case when a and b have opposite sign Gilioli and Treves use a variant of Treves' general method of concatenations to prove that P fails to be locally solvable at the origin if and only if $\frac{c}{a-b}$ is an integer congruent to 0 or 1 mod $(k+1)$. Since P is an operator of Grushin's type. of index 0, it follows that the above conditions for local solvability are also the conditions for hypoellipticity. Thus, $\frac{c}{a-b}$ is an integer congruent to 0 or 1 mod $(k+1)$ precisely when 0 is an eigenvalue of one of the partial Fourier transformed operators.

$$
\begin{aligned}
P(t,D_t,1) &= \left(\frac{\partial}{\partial t} + at^k\right)\left(\frac{\partial}{\partial t} + bt^k\right) - ct^{k-1} \\
P(t,D_t,-1) &= \left(\frac{\partial}{\partial t} - at^k\right)\left(\frac{\partial}{\partial t} - bt^k\right) + ct^{k-1}
\end{aligned}
\tag{1.2}
$$

In the case $k = 1$, we can restate this as: c is an eigenvalue of $\left(\frac{\partial}{\partial t} + at\right)\left(\frac{\partial}{\partial t} + bt\right)$ or $-c$ is an eigenvalue of $\left(\frac{\partial}{\partial t} - at\right)\left(\frac{\partial}{\partial t} - bt\right)$ if and only if $\frac{c}{a-b}$ is an integer. That is, the method of concatenations provides a way to actually compute the eigenvalues of the partial Fourier transformed operators.

The operators treated by Grushin share two properties which make them rather special: 1) They are written in a special coordinate representation with certain variables singled out for the purpose of partial Fourier transformation.

2) They satisfy certain quasi-homogeneity properties with respect to this coordinate system.

Treves in [25] treats a class of operators still written in a distinguished coordinate system, but <u>not</u> satisfying quasi-homogeneity properties. Specifically, he treats abstract second-order evolution operators of the type

$$(1.3) \qquad P = \left(\frac{\partial}{\partial t} - a(t,A)A\right)\left(\frac{\partial}{\partial t} - b(t,A)A\right) - c(t,A)A .$$

Here A is an <u>unbounded</u>, densely defined, self-adjoint positive-definite linear operator on a Hilbert space H, with <u>bounded</u> inverse A^{-1}. The expressions $a(t,A)$, $b(t,A)$, $c(t,A)$ are power series in non-negative powers of A^{-1} with coeeficients C^{∞} functions of t. These power series, as well as all their t-derivatives, are assumed to converge in $B(H,H)$, the space of bounded linear operators on H. Notice that the operators of type (1.3) are "non quasi-homogeneous" generalizations of operators of type (1.1). Treves makes the restriction that $a_o(t)$ and $b_o(t)$, the leading coefficients of $a(t,A)$, $b(t,A)$ respectively, vanish at $t = 0$, but that $a_o'(0) \neq 0$ and $b_o'(0) \neq 0$. In terms of (1.1) this corresponds to the case of $k = 1$. Having made the natural definitions of local solvability and hypoellipticity for abstract operators of type (1.3). Treves goes on to show that if $\text{Re}\, a_o'(0) > 0$ and $\text{Re}\, b_o'(0) > 0$ then P is hypoelliptic but not locally solvable at $t = 0$; and that if $\text{Re}\, a_o'(0) < 0$ and $\text{Re}\, b_o'(0) < 0$ then P is locally solvable but not hypoelliptic at $t = 0$.. If $\text{Re}\, a_o'(0)$ and $\text{Re}\, b_o'(0)$ have opposite sign he shows that P is locally solvable at $t = 0$ if and only if P is hypoelliptic at $t = 0$, and, furthermore, quite in analogy with (1.1) he derives, by means of his general method of concatenations, a set of <u>discrete</u> conditions such that P is locally solvable (and hypoelliptic) at $t = 0$ if and only if none of the discrete conditions hold. More precisely, he obtains a sequence $c^j(A)$, $j = 0,\ldots$ of <u>formal</u> (i.e. not necessarily convergent) power series in non-negative powers of A^{-1} with coefficients in $\mathbb{C}$, and proves that local solvability and hypoellipticity at $t = 0$ are equivalent to the condition that none of the formal power series $c^j(A)$ has all its coefficients equal to zero.

Results have also been obtained by Sjostrand ([24]) and Boutet de Mouvel and Treves ([3] and [4]) in a fairly general setting, generalizing that of (1.1) with $k = 1$, where there is no distinguished set of coordinates with which to take the partial Fourier transform of P. We shall discuss these results as they appear in [3] and [4]. Boutet de Mouvel and Treves treat a sharp form of hypoellipticity (and local solvability), namely hypoellipticity with loss of 1 derivative. This is the strongest hypoellipticity condition which the lower order part of P (i.e., anything other than the <u>principal</u> symbol) could possibly influence. They show that this hypoellipticity condition is microlocalizable, i.e., can be lifted to the context of the cotangent bundle and there be reduced to corresponding conditions on conic neighborhoods. They introduce two bilinear forms along each fiber of the conormal bundle of the characteristic variety Σ of P, one symmetric, coming from the principal symbol of P, and one anti-symmetric, coming from the canonical symplectic forms on the cotangent space. Assuming a condition akin to the condition for (1.3) that $\operatorname{Re} a'(0)$ and $\operatorname{Re} b'(0)$ have opposite sign, they use another variant of the method of concatenations to derive a set of discrete conditions such that P is hypoelliptic with loss of one derivative (and locally solvable with loss of one derivative) if and only if none of the discrete conditions hold. In fact, if $2n$ is the codimension of Σ in the cotangent space, then at each point (x,ξ) of Σ there is an n-parameter family (with integral parameters) of conditions expressed in terms of the symmetric and anti-symmetric forms defined at (x,ξ). For operators of type (1.3) modelled on (1.1) with k not necessarily equal to 1, the analogue of hypoellipticity with loss of one derivative is hypoellipticity with loss of $\frac{2k}{k+1}$ derivatives. The situation is treated by Gilioli ([11]), who derives necessary and sufficient conditions. Again, these take the form of a discrete family of conditions to be avoided.

Boutet de Monvel ([2]) describes a symbolic calculus

which allows the construction of parametrices for operators of the type treated in [4] as well as operators of "heat-equation" type. Folland ([17]) and Folland and Stein ([9],[10]) have constructed explicit parametrices, acting on the L^p spaces $(p>1)$ and on the Hölder spaces, for $\Box_b$, the Laplacian of the tangential Cauchy-Riemann complex.

Many of these results can be viewed in a unified manner, both in the cases when discrete conditions arise and in the cases in which they do not. The unifying theme is that of eigenvalue asymptotics. In this framework the method of concatenations appears as a procedure for computing eigenvalues, a generalization of the physicist's procedure for computing the eigenvalues of the quantum mechanical harmonic oscillator by means of commutation relations. (See, for example, [21], Chap. XII.). In §2, we shall show how the n-parameter family of conditions for hypo-ellipticity (and local solvability) with loss of 1 derivative may be viewed as eigenvalue conditions. To each point (x,ξ) of Σ, the characteristic variety, we associate an invariantly defined test-operator, $\tilde{P}_{(x,\xi)}$, constructed from the principal symbol and subprincipal symbol of P. $\tilde{P}_{(x,\xi)}$ is a differential operator with polynomial coefficients acting on $L^2(R^n)$, where $2n$ equals the codimension of Σ. More precisely, it is the unitary equivalence class of $\tilde{P}_{(x,\xi)}$ which is given. We shall see that hypoellipticity of P with loss of 1 derivative is equivalent to each $\widetilde{P}_{(x,\xi)}$ being injective as a Hilbert space operator, i.e., is equivalent to 0 not being an eigenvalue of $\widetilde{P}_{(x,\xi)}$. Each of our test-operators has an index, analoguous to Grushin's index for his partial Fourier transformed operators, and we shall see that this index is 0 precisely when the condition of Boutet de Monvel and Treves, akin to the condition that $\operatorname{Re} a_0'(0)$ and $\operatorname{Re} b_0'(0)$ have opposite sign, holds. In this case $\widetilde{P}_{(x,\xi)}$ has an n-parameter family (with integral parameter) of eigenvalues. We shall compute these eigenvalues in the general case when P has a scalar principal symbol by the method of commutation relations. When P has a real principal symbol the test-operators always consist of n independent harmonic oscillators, occurring with various "weights", plus a constant term, the n "weights" and the constant term varying with the point $(x,\xi) \in \Sigma$. (Our use of the term "weights" will be clear from context). In the case of a real principal symbol we shall also examine the eigenvalues from the standpoint of Maslov asymptotics.

In §3 we shall use the results of §2 to derive necessary and sufficient conditions for hypoellipticity with loss of 1 derivative for the various Laplacians $\Delta_i : E^i \longrightarrow E^i$ associated to a Poincare complex $\{P, E^i\}$ of first-order operators with simple characteristics. In particular, this includes the case of Kohn's tangential Cauchy-Riemann complex $\bar{\partial}_b$. We will see that under the assumption of a non-degenerate Levi-form the necessary and sufficient conditions for hypoellipticity with loss of 1 derivative of Δ_k agrees with the conditions (see [15], [23]) for 1/2-subellipticity of the complex $\{P,E^i\}$ at the k-th position. We note that the principal symbol of Δ_k is real, so that, as pointed out earlier, the test-operator at each point of the characteristic variety consists, apart from a constant term, of n independent harmonic oscillators. It terms out that the n weights at each point $(x,\xi) \in \Sigma$ consist, essentially, of the absolute values of the eigenvalues of the Levi-form $\frac{1}{\sqrt{-1}}\{p_i,\bar{p}_j\}_{(x,\xi)}$. We remark also, although we shall not develop this point, that the test-operators constructed in this paper seem closely related to the test-complexes constructed in [22] and [23] for the study of general first-order complexes.

In §4 we introduce natural notions of "eigenvalue" and "asymptotic eigenvalue" for abstract operators P of type (1.3). We show that the formal power series $-c^j(A)A$, $j = 0,1,\ldots$ are precisely the asymptotic eigenvalues of P. Thus, Treves' condition for hypoellipticity of P. . No eigenvalue of P has its asymptotic expansion identically equal to 0 . If we were to form the analogue of the test-operators in the context of (1.3) we would see that the exact eigenvalues of the test-operators are the leading terms $-c_0^j(A)A$ of the asymptotic eigenvalues $-c^j(A)A$. Consequently, the condition for hypoellipticity of P with loss of 1 derivative, the "strongest" hypoellipticity condition, may be stated as: No eigenvalue of P has the leading term of its asymptotic expansion equal to 0.

The Appendix contains a remark on "test-operators" in the case of simple characteristics.

We point out that rather striking similarities exist between the theory associated with the test-operators and the Kirillov theory of representations of nilpotent Lie groups ([17], [1]). The analogy with the Kirillov theory suggests that it may be possible to construct a left (right) parametrix for P by some sort of process which would involve taking a left (right) inverse for $\widetilde{P}_{(x,\xi)}$ at each $(x,\xi) \in \Sigma$ and then forming some type of "direct integral" (or inverse partial Fourier transform) over Σ. However, it is too early to say more about this here. A somewhat different link between nilpotent Lie groups and "general" differential operators is brought out by the work of Follard and Stein ([9], [10]) mentioned earlier, where the Heisenberg group, the simplest non-abelian nilpotent Lie group, is used to "approximate" a general strongly pseudo-convex manifold. In fact (see §3), the test-operator for the $\bar{\partial}_b$ Laplacian occurs in ([9], [10]) in a form explicitly linked to the representation theory of the Heisenberg group.

In conclusion, we feel that the circle of ideas centering about the test-operators will be applicable in a more general setting than that treated here. The work presented here should be regarded only as preliminary steps in this direction.

§2. Hypoellipticity with loss of one derivative

§2.1 Introduction and statement of theorem

In this section we show how the conditions of Boutet de Monvel and Treves ([3] and [4]) for hypoellipticity with loss of one derivative may be naturally interpreted via the eigenvalue standpoint.

Let X be a C^{∞} manifold of dimension k and let

$$(2.1) \qquad P(x,D) = p(x,D)I + Q(x,D)$$

be a determined (i.e. square $N \times N$) system of pseudo-differential operators. Let m be the order of P. We assume that the principal symbol $p(x,\xi)$ on $T^*X\setminus 0$ (C^{∞} and positive-homogeneous of degree m in ξ) is scalar, i.e., complex-valued. The $N \times N$ identity matrix is denoted by I. The remaining term, $Q(x,D)$, is an $N \times N$ system of pseudo-differential operators of lower order, i.e., of order $m-1$ or less. We remark that allowing $Q(x,D)$ to be matrix valued rather than scalar valued is not due purely to a desire for utmost generality, but, rather, is quite natural. Indeed, if $P(x,D)$ is, for example, the Laplacian (at some position) of a complex of differential operators, for example the tangential Cauchy-Riemann complex $\bar{\partial}_b$, then although the principal symbol is scalar the lower order part may be matrix-valued. In §3 we shall treat in detail from the viewpoint of this paper the $\bar{\partial}_b$ Laplacian. [In fact we shall treat an arbitrary Poincaré complex with simple characteristics.]

We need to make further assumptions about p. We let $\omega = \sum_{i=1}^{k} d\xi_i \wedge dx_i$ be the standard symplectic form on $T^*X \setminus 0$. Let Σ denote the characteristic variety of P, i.e.,

<u>Definition 2.1</u> $\Sigma = \{(x,\xi) \in T^*X\setminus 0 \mid p(x,\xi) = 0\}$.

Since p is homogeneous in ξ, Σ is conic, that is, if $(x,\xi) \in \Sigma$ then $(x,r\xi) \in \Sigma$ for every $r \in \mathbb{R}^+$. We assume:

(2.2) Σ is a C^∞ submanifold of $T^*X\setminus 0$.

(2.3) The pull-back of ω to Σ via the injection $\Sigma \xrightarrow{i} T^*X$ is non-degenerate. This of course, makes (Σ, ω) a symplectic manifold.

<u>Remark:</u> In the case of $P = \bar{\partial}_b$-Laplacian at some position, (2.3) is just the condition that the Levi-form be non-degenerate. [see §3].

Condition (2.3) implies that the dimension of Σ is even. Hence, since dim $T^*X\setminus 0$ is even, the codimension of Σ in $T^*X\setminus 0$ is even.

Definition 2.2 $n = 1/2 \text{ codim } \Sigma$

We need one more condition.

(2.4) p vanishes exactly to order 2 on Σ.

This condition can be expressed: $\dfrac{|\xi|^{-m+2}|p(x,\xi)|}{d(x,\xi)^2}$

is locally bounded both above and below in $T^*X\backslash 0$. Here $d(x,\xi)$ denotes the distance from (x,ξ) to Σ. Condition (2.4) may be viewed as akin to the condition in §1. that we take $k = 1$ for the operator (1.1).

Remark 2.2A Boutet de Monvel and Treves insist upon one further condition, namely that if $n = 1$ that the winding number of $p(x,\xi)$ about Σ is 0. This condition is essentially the same as the condition in §1. that for the operator (1.3) $\text{Re } a_0'(0)$ and $\text{Re } b_0'(0)$ have opposite sign. It is necessary in order for discrete conditions to arise, and in order for the method of concatenations to be applicable. However, it does not appear to be a necessary condition for an eigenvalue criterion to hold. We shall discuss this in some detail later when we deal with the index of our test-operators.

We need the following definition

Definition 2.3. P is hypoelliptic with loss of one derivative if for every open subset U of X, for every

$s \in \mathbb{R}$, and for every distribution u in U

$$P(x,D)\, u \in H^{s}_{loc}(U) \Longrightarrow u \in H^{s+m-1}_{loc}(U)$$

where H^{s}_{loc} denotes the standard (localized) Sobolev spaces.

We point out that hypoellipticity with loss of one derivative is the strongest hypoellipticity for which the lower-order part of P can play a role. That is, the condition of hypoellipticity with loss of $(1-\varepsilon)$ derivatives, where $e > 0$, i.e., $Pu \in H^{s}_{loc}(U) \Longrightarrow u \in H^{s+m-(1-\varepsilon)}_{loc}(U)$, depends purely on the principal symbol p of P and not at all on the lower order part of P.

We shall show how the necessary and sufficient conditions for hypoellipticity with loss of one derivative can be stated as an eigenvalue criterion for certain test-operators. Before stating our theorem, we need to discuss some preliminary notions.

Let $N(\Sigma)$ denote the conormal bundle of Σ in $T^{*}X\backslash 0$. We shall see that the non-degeneracy assumption (2.3) implies that for each point $(x,\xi) \in \Sigma$ the fiber $N(\Sigma)_{(x,\xi)}$ may be made into a symplectic vector space of dim $2n$ [or, alternatively, we may prefer the viewpoint of a symplectic manifold of dim $2n$]. We shall consider $\tilde{p}_{(x,\xi)}$, a symmetric form defined on $N(\Sigma)_{(x,\xi)}$ via the Hessian of p. We shall consider the quadratic polynomial on $N(\Sigma)_{(x,\xi)}$ given by $v \longmapsto \tilde{p}_{(x,\xi)}(v,v)$, and shall show how to associate to $\tilde{p}_{(x,\xi)}(v,v)$, in an invariant way, a unitary equivalence class of operators $\tilde{P}_{(x,\xi)}$ on a Hilbert space $L^2(V_{(x,\xi)})$ invariantly associated to $N(\Sigma)_{(x,\xi)}$ up to unitary equivalence. In fact, $V_{(x,\xi)}$ can be taken as any Lagrangian subspace of the 2n-dimensional symplectic vector space $N(\Sigma)_{(x,\xi)}$ i.e., an n-dimensional subspace of $N(\Sigma)_{(x,\xi)}$ which is self-annihilating with respect to the symplectic form on $N(\Sigma)_{(x,\xi)}$. Heuristically, $V_{(x,\xi)}$ is "the space" having $N(\Sigma)_{(x,\xi)}$ as its cotangent space. Since we give $\tilde{P}_{(x,\xi)}$ uniquely up to unitary equivalence it follows, in particular, that the eigenvalues of $\tilde{P}_{(x,\xi)}$ are well-defined.

On the other hand we shall show that we can invariantly define the notion of $\sigma_{sub}(P)|_\Sigma$, the restriction to Σ of the sub-principal part of P. Our theorem will be

Theorem 2.4 Let P be of type (2.1) and suppose that conditions (2.2) - (2.4) are satisfied. Then P is hypoelliptic with loss of one derivative $\Longleftrightarrow$ for every $(x,\xi) \in \Sigma$, $-\sigma_{sub}(P)|_{(x,\xi)}$ is not an eigenvalue of $\tilde{P}_{(x,\xi)}$. That is, P is hypoelliptic with loss of one derivative if and only if for every $(x,\xi) \in \Sigma$

$\tilde{P}_{(x,\xi)} + \sigma_{sub}(P)|_{(x,\xi)}$ does not have 0 as an eigenvalue.

Remark: We have stated the theorem as if $Q(x,D)$ were necessarily scalar rather than matrix-valued. In the latter case $\sigma_{sub}(P)|_{(x,\xi)}$ is an NxN matrix rather than a complex number. The hypoellipticity condition then becomes

For every $(x,\xi) \in \Sigma$

$$\tilde{P}_{(x,\xi)} I_{NxN} + \sigma_{sub}(P)|_{(x,\xi)} : \underbrace{L^2(V_{(x,\xi)}) \oplus \cdots \oplus L^2(V_{(x,\xi)})}_{N \text{ factors}}$$

$$\longrightarrow \underbrace{L^2(V_{(x,\xi)}) \oplus \cdots \oplus L^2(V_{(x,\xi)})}_{N \text{ factors}}$$

does not have 0 as an eigenvalue.

By writing $\sigma_{sub}(P)|_{(x,\xi)}$ in Jordan canonical form it is easy to show that the preceding condition is equivalent to the following:

Let $\mu_1|_{(x,\xi)}, \ldots, \mu_N|_{(x,\xi)}$ be the eigenvalues (counting multiplicities) of $\sigma_{sub}(P)_{(x,\xi)}$. For every $(x,\xi) \in \Sigma$ and for every $i = 1,\ldots,N$, $\tilde{P}_{(x,\xi)} + \mu_i|_{(x,\xi)}$ does <u>not</u> have 0 as an eigenvalue. That is, for every $(x,\xi) \in \Sigma$ and for every $i = 1,\ldots,N$, $-\mu_i|_{(x,\xi)}$ is <u>not</u> an eigenvalue of $\tilde{P}_{(x,\xi)}$. (See, §2.5, Lemma 2.47).

We should mention at this point that the test-operators $\tilde{P}_{(x,\xi)} + \sigma_{sub}(P)|_{(x,\xi)}$ are homogeneous (of degree m-1) with respect to the $\mathbb{R}^+$ action on Σ, in the sense, for example, that the eigenvalues $\lambda_j|_{(x,\rho\xi)}$ of $\tilde{P}_{(x,\rho\xi)} + \sigma_{sub}(P)|_{(x,\rho\xi)}$ are $\rho^{m-1}\lambda_j|_{(x,\rho\xi)}$.

Hence, if we fix a point $(x_\circ, \xi_\circ)$ and consider the asymptotic expansion of these eigenvalues along the ray determined by $(x_\circ, \xi_\circ)$, i.e., the asymptotic expansion in, essentially, non-negative integral powers of ρ^{-1} of the eigenvalue $\lambda_j|_{(x_\circ, \rho\xi_\circ)}$, then this expansion has 0 as every coefficient after the first. This would correspond, in the case of the operators (1.3) of §1. to all the formal power series $c^j(A)$ having all their coefficients, except possibly their leading coefficients, equal to 0. Thus, for such $c^j(A)$, having their leading coefficients vanish is the same as having <u>all</u> their coefficients vanish.

We point out here that the set-up consisting of a fibering $N(\Sigma) \longrightarrow \Sigma$ each fiber being a symplectic manifold, over a space Σ with an $\mathbb{R}^+$ action (in this case given by $\langle\rho, (x,\xi)\rangle \longmapsto (x, \rho\xi)$, for $\rho \in \mathbb{R}^+$ and $(x,\xi) \in \Sigma)$, together with the assignment to each point $(x,\xi) \in \Sigma$ of a differential operator $P_{(x,\xi)}$ on $L^2(V_{(x,\xi)})$ (up to unitary equivalence independent of the choice of "polarization" $V_{(x,\xi)}$ of $N(\Sigma)_{(x,\xi)}$) seems a natural context in which to treat Maslov-type asymptotics ([19], [20]). Notice, however, that there are two differences between our context and Maslov's. First, Maslov deals with one fixed symplectic manifold, namely T^*Y for some space Y, rather than with a family of symplectic manifolds. Second, Maslov studies asymptotic behavior in $\frac{1}{h}$ as "Plank's constant" h goes to 0, whereas we study asymptotic behavior in ρ as $\rho \longrightarrow +\infty$. Maslov also treats the case where $\frac{1}{h}$ is replaced by an unbounded operator on a Hilbert space. This is related to the material we shall treat in §4. We shall return to the topic of Maslov asymptotics in §2.5. We shall show, for example, how, at least in the case when p is real-valued, the eigenvalues of $\tilde{P}_{(x,\xi)}$ are picked out by appropriate Lagrangian submanifolds lying in the symplectic manifold $N(\Sigma)_{(x,\xi)}$.

We turn now to the details of Theorem 2.4.

Lemma 2.5. The assumption (2.3) that ω pulled back to Σ is non-degenerate allows us to intrinsically define a non-degenerate anti-symmetric form denoted $\omega_{(x,\xi)}$ on the vector space $N(\Sigma)_{(x,\xi)}$. We can thus view each $N(\Sigma)_{(x,\xi)}$ as a symplectic vector space.

We first need to prove

Sublemma 2.6. Let $\omega: E \times E \longrightarrow \mathbb{R}$ be a non-degenerate anti-symmetric form on the finite dimensional vector space E. Let F be any subspace of E, and let $F^{\perp}$ denote orthogonal complement with respect to ω. Then the following conditions are equivalent:

(i) $\omega|_F$ is non-degenerate.

(ii) $F \cap F^{\perp} = \{0\}$

(iii) $\omega|_{F^{\perp}}$ is non-degenerate

Pf:

That $\omega|_F$ is non-degenerate means precisely that $F \cap F^{\perp} = \{0\}$ .. That $w|_{F^{\perp}}$ is non-degenerate means that $F^{\perp} \cap (F^{\perp})^{\perp} = \{0\}$. So, to prove the equivalence of (i), (ii), (iii) it suffices to show that $F = (F^{\perp})^{\perp}$. But $\dim E = \dim F + \dim F^{\perp}$, for $\omega\colon E \times E \longrightarrow \mathbb{R}$ non-degenerate $\Longrightarrow F^{\perp}$ may be identified with the annihilator of F in E^*, the dual space of E. The same argument yields that $\dim E = \dim F^{\perp} + \dim (F^{\perp})^{\perp}$. Therefore, $\dim F = \dim (F^{\perp})^{\perp}$. But, clearly, $F \subset (F^{\perp})^{\perp}$, so the equality of dimensions $\Longrightarrow F = (F^{\perp})^{\perp}$.

QED

Since $\omega|_{F^{\perp}}$ is non-degenerate, and since $\omega\colon E \times E$ ----- $\mathbb{R}$, being non-degenerate, gives us a canonical pairing $\theta\colon E \longrightarrow E^*$ [given by $\omega(v,w) = [\theta(w)](v) \quad \forall v,w \in E$] we get, by transferring ω via θ, a non-degenerate anti-symmetric form $\omega\colon \theta(F^{\perp}) \times \theta(F^{\perp}) \longrightarrow \mathbb{R}$. But, as we pointed out in the proof of Sublemma 2.6, $\theta(F^{\perp})$ = annihilator of F in E^*. Now taking $E = T\,(T^*X\backslash 0)_{(x,\xi)}$ and $F = T(\Sigma)_{(x,\xi)}$ we get precisely that the annihilator of F in E^* equals $N(\Sigma)_{(x,\xi)}$. This proves Lemma 2.5.

QED

<u>Remark 2.7</u>. Clearly, if we represent $v,w \in N(\Sigma)_{(x,\xi)}$ as $df_{(x,\xi)}$, $dg_{(x,\xi)}$ where f, g are real-valued C^∞ functions defined in a neighborhood of (x,ξ) in $T^*X\setminus 0$, then $\omega_{(x,\xi)}(v,w) = \omega_{(x,\xi)}(df_{(x,\xi)},dg_{(x,\xi)}) = \omega(H_f,H_g)_{(x,\xi)} = \{f,g\}_{(x,\xi)}$. Here H_f, H_g denote the Hamiltonian vector fields corresponding to f, g, and $\{\ ,\ \}$ denotes Poisson bracket.

We state next an alternate way of viewing the result of Lemma 2.5 which will be useful when we deal with the Maslov asymptotics. Rather than view $(N(\Sigma)_{(x,\xi)}, \omega_{(x,\xi)})$ as a symplectic <u>vector space</u> of dimension $2n$, we may view it as a symplectic <u>manifold</u> of dimension $2n$. Indeed, since there is a standard identification of a vector space V with its tangent space $T(V)_v$ at any point v (given by $w \in V \longmapsto \psi:[-1,1] \longrightarrow V$ where $\psi: t \longmapsto v+tw$) we may view the anti-symmetric non-degenerate form $_{(x,\xi)}$ on $N(\Sigma)_{(x,\xi)}$ considered as a vector space as a non-degenerate 2-form on $N(\Sigma)_{(x,\xi)}$ considered as a manifold. Since this 2-form $_{(x,\xi)}$ has constant coefficients with respect to any linear coordinate system it is, in particular, closed. Hence, $(N(\Sigma))_{(x,\xi)}, {}_{(x,\xi)})$ is indeed a symplectic manifold.

<u>Lemma 2.8.</u> The assumption (2.4) that p vanishes exactly to order 2 on Σ allows us to intrinsically define a non-degenerate symmetric form $\tilde{p}_{(x,\xi)}$ on the vector space $N(\Sigma)_{(x,\xi)}$. In case p is real-valued $\tilde{p}_{(x,\xi)}$ is strictly positive or negative definite.

Pf:

Since p vanishes to at least second order on Σ we know that $dp|_{(x,\xi)} = 0$, so the Hessian

$$\text{Hess } p|_{(x,\xi)} : T(T^*X\backslash 0)|_{(x,\xi)} \times T(T^*X\backslash 0)|_{(x,\xi)} \longrightarrow \mathbb{R}$$

can be intrinsically defined as follows: For any $w_1, w_2 \in T(T^*X\backslash 0)_{(x,\xi)}$ choose vector fields X_1, X_2 on $T^*X\backslash 0$, defined in a neighborhood of (x,ξ), such that $X_i|_{(x,\xi)} = w_i$, $i = 1,2$. Then $\text{Hess } p|_{(x,\xi)}(w_1,w_2) = X_1\,(X_2(p))|_{(x,\xi)}$. To see that this definition is well-defined and that $\text{Hess } p|_{(x,\xi)}$ thus defined is symmetric we proceed as follows: It is clear that $X_1\,(X_2(p))|_{(x,\xi)}$ depends on X only to the extent that one needs to know $X_1|_{(x,\xi)}$. But this equals w_1. Likewise, insofar as it depends on X_2, $X_2\,(X_1(p))|_{(x,\xi)}$ depends only on $X_2|_{(x,\xi)}$. But this equals w_2. Thus, both to show well-definedness and symmetry, it suffices to show that $X_1\,(X_2(p))|_{(x,\xi)} = X_2\,(X_1(p))|_{(x,\xi)}$. But the difference equals $[X_1,X_2]\,(p)|_{(x,\xi)}$ Since $[X_1,X_2]$ is also a vector field and since $dp|_{(x,\xi)} = 0$ we see that $[X_1,X_2]\,(p)|_{(x,\xi)} = 0$.

The following analoguous argument shows that, by using the additional fact that $dp|_\Sigma \equiv 0$, i.e. that $dp = 0$ at every point of the manifold Σ, we can show that

$$\text{(2.5)} \quad \text{Hess } p|_{(x,\xi)}\,(w_1,w_2) = 0 \qquad \text{for every } w_2 \in T(T^*X\backslash 0)_{(x,\xi)} \quad \text{if } w_1 \in T(\Sigma)_{(x,\xi)}.$$

Indeed, choose vector fields X_1, X_2 such that $X_i|_{(x,\xi)} = w_1$. Since $dp|_\Sigma \equiv 0$ we know that $X_2(p)|_\Sigma \equiv 0$. Since w_1 is tangential to Σ it follows that $w_1\,(X_2(p)) = 0$, which proves (2.5).

From (2.5) we see that $\text{Hess}\, p|_{(x,\xi)}$ induces a symmetric bilinear form, which we continue to denote $\text{Hess}\, p|_{(x,\xi)}$

$$(2.6) \qquad \text{Hess}\, p|_{(x,\xi)} : \mathcal{N}(\Sigma)_{(x,\xi)} \times \mathcal{N}(\Sigma)_{(x,\xi)} \longrightarrow \mathbb{R}$$

where $\mathcal{N}(\Sigma) \equiv T(T^*X\backslash 0)/T(\Sigma)$, i.e., $N(\Sigma)$ is the <u>normal</u> as opposed to <u>conormal</u> bundle of Σ. The bounded below assumption contained in condition (2.4) can easily be shown to imply that if p vanishes exactly to second order on Σ then

(2.7) $\text{Hess}\, p|_{(x,\xi)}([v],[v]) \neq 0$ unless $[v] = 0$, where v is any element of $T(T^*X\backslash 0)|_{(x,\xi)}$ and $[\ ]$ denotes coset in $\mathcal{N}(\Sigma)_{(x,\xi)}$. [In particular $\text{Hess}\, p|_{(x,\xi)}$ in (2.6) is non-degenerate].

If p is real-valued then (2.7) implies that the symmetric form $\text{Hess}\, p|_{(x,\xi)}$ in (2.6) is strictly definite (either positive or negative). Indeed, if $\mathcal{N}(\Sigma)_{(x,\xi)}$ is one-dimensional the result is immediate. If $\dim \mathcal{N}(\Sigma)_{(x,\xi)} > 1$, then $\mathcal{N}(\Sigma)_{(x,\xi)} - \{0\}$ is connected, and so the image of $\mathcal{N}(\Sigma)_{(x,\xi)} - \{0\}$ under the map $[v] \longmapsto \text{Hess}\, p|_{(x,\xi)}([v],[v])$ is a connected subset of $\mathbb{R}$ which, according to (2.7), does not contain $\{0\}$.

The result then follows. (Of course we know $\dim \mathcal{N}(\Sigma)_{(x,\xi)} > 1$ in our case since codim Σ is even.)

Remark 2.9 An analoguous definiteness condition can be derived even when p is not real-valued if, in case $n = 1$, we make the additional assumption that the winding number of $p(x,\xi)$ about Σ equals 0. (See Remark 2.2A). Indeed, it is pointed out in [4] that Sjostrand in [24] proves, assuming the above additional condition holds if $n = 1$, that there exists $z \in \mathbb{C}$ such that $\operatorname{Re} z(\operatorname{Hess} p|_{(x,\xi)}([v],[v]))$ is greater than 0 wherever $[v] \neq 0$.

We have defined a symmetric non-degenerate form on $\mathcal{N}(\Sigma)_{(x,\xi)}$. We shall show how to transfer it to a form on $N(\Sigma)_{(x,\xi)}$ via the pairing $\theta : T(T^*X\backslash 0)_{(x,\xi)} \longrightarrow T^*(T^*X\backslash 0)_{(x,\xi)}$ given by ω and discussed in the proof of Lemma 2.5. We saw in the proof of Lemma 2.5 that $\theta\,(T(\Sigma)^{\perp}_{(x,\xi)}) = N(\Sigma)_{(x,\xi)}$. Let $[\ ] : T(\Sigma)^{\perp}_{(x,\xi)} \longrightarrow \mathcal{N}(\Sigma)_{(x,\xi)}$ be given by $v \longmapsto [v]$. Since $\dim T(\Sigma)^{\perp}_{(x,\xi)} = \dim \mathcal{N}(\Sigma)_{(x,\xi)}$, $[\ ]$ will be bijective if we can show it is injective. But injectivity follows from the fact that $T(\Sigma)_{(x,\xi)} \cap T(\Sigma)^{\perp}_{(x,\xi)} = \{0\}$, which in turn follows from the fact that $\omega|_{T(\Sigma)_{(x,\xi)}}$ is non-degenerate.

Thus, we may transfer $\operatorname{Hess} p|_{(x,\xi)}$ to a form on $N(\Sigma)_{(x,\xi)}$ via the bijection $[\ ] \circ \theta^{-1} : N(\Sigma)_{(x,\xi)} \longrightarrow \mathcal{N}(\Sigma)_{(x,\xi)}$. Hence, we may complete the proof of Lemma 2.8 by defining

$\tilde{p}_{(x,\xi)} : N(\Sigma)_{(x,\xi)} \times N(\Sigma)_{(x,\xi)} \longrightarrow \mathbb{C}$ as follows:

Definition 2.10 $\quad \tilde{p}_{(x,\xi)} : (v_1, v_2) \longmapsto$

$$\tfrac{1}{2} \text{ Hess } p|_{(x,\xi)} ([\theta^{-1}(v_1)], [\theta^{-1}(v_2)]) \quad .$$

Remark 2.11 Using the same notation as in Remark 2.7 we can see that

$$\tilde{p}_{(x,\xi)} (df_{(x,\xi)}, dg_{(x,\xi)}) = \frac{1}{2} \{f, \{g,p\}\}|_{(x,\xi)} \; .$$

In fact, $\tilde{p}_{(x,\xi)} (df_{(x,\xi)}, dg_{(x,\xi)}) = \frac{1}{2} \text{ Hess } p|_{(x,\xi)}$

$$(H_f|_{(x,\xi)}, H_g|_{(x,\xi)}) = \frac{1}{2} H_f (H_g(p))|_{(x,\xi)} =$$

$$\frac{1}{2} \{f, H_g(p)\}|_{(x,\xi)} = \frac{1}{2} \{f, \{g,p\}\}|_{(x,\xi)} .$$

In particular, if $u_1, \ldots, u_{2n}$ are a local set of parameters for Σ at $(x,\xi) \in \Sigma$, i.e. , Σ is given locally at (x,ξ) by $u_1 = \cdots = u_{2n} = 0$, and $du_1, \ldots, du_{2n}$ are linearly independent at (x,ξ) , then $\omega_{(x,\xi)} (du_i, du_j) = \{u_i u_j\}|_{(x,\xi)}$ and $\tilde{p}_{(x,\xi)} (du_i, du_j) = \frac{1}{2} \{u_i, \{u_j, p\}\}|_{(x,\xi)}$.

§2.2 <u>The test-operators</u>

Our next object is to associate to the polynomial $\tilde{p}_{(x,\xi)}(v,v)$, in invariant fashion, a unitary equivalence class of operators $\tilde{P}_{(x,\xi)}$ on a Hilbert space. The idea, basically, is this. Choose canonical linear coordinates $s_1,\ldots,s_n$, $t_1,\ldots,t_n$ on $N(\Sigma)_{(x,\xi)}$, i.e., linear coordinates such that $\omega_{(x,\xi)} = \Sigma\, dt_i \wedge ds_i$. In these coordinates $\tilde{p}_{(x,\xi)}$ is a polynomial homogeneous of degree two in the s_i's and t_i's. Then we would like to get a differential operator $\tilde{P}_{(x,\xi)}$ on s-space by replacing s_i in $\tilde{p}_{(x,\xi)}$ by multiplication by s_i and t_i by $D_{s_i} = \frac{1}{\sqrt{-1}} \frac{\partial}{\partial s_i}$. But there are two problems. First, does $s_i t_i$ go over to $s_i D_{s_i}$ or $D_{s_i} s_i$ or something still different, for example $\frac{1}{2}(s_i D_{s_i} + D_{s_i} s_i)$? Second, what happens if we choose a different set of canonical linear coordinates? Is the resulting differential operator on $\mathbb{R}^n$ unitarily equivalent to the differential operator on $\mathbb{R}^n$ obtained via the first choice of canonical linear coordinates? Below we shall see how to set up such a correspondence in invariant fashion.

Consider 2n-dimensional "phase-space" $\mathbb{R}^{2n}$, with <u>fixed</u> coordinates $s_1,\ldots,s_n$, $t_1,\ldots,t_n$, and let G denote the group of symplectic transformations on $\mathbb{R}^{2n}$ (i.e., linear canonical transformations). Let $S^2\mathbb{R}^{2n}$ be the vector space (over $\mathbb{C}$) of all (constant-coefficient) polynomials on $\mathbb{R}^{2n}$

homogeneous of degree 2. Let $\mathrm{Diff}^2(\mathbb{R}^n)$ denote the vector space (over $\mathbb{C}$) of all linear differential operators on $\mathbb{R}^n$ of the form

(2.8)
$$\Sigma\; a_{ij} D_{s_i} D_{s_j} + \Sigma\; b_{ij} s_i s_j + \Sigma c_{ij} s_i D_{s_j} + \Sigma d_{ij} D_{s_j} s_i + \Sigma e_i D_{s_i} + \Sigma f_i s_i + g$$

[There is no reason to automatically **write** $D_{s_i} s_i$ **as** $s_i D_{s_i} + \frac{1}{\sqrt{-1}}$, and we shall **in fact see that a different** normal form is preferable.]

<u>Remarks</u>: <u>1</u>. If $p \in \mathrm{Diff}^2(\mathbb{R}^n)$ then it is clear that the coefficients e_i, f_i are uniquely determined, independent of the choice of representation of type (2.8) . (Remember, the coordinates $s_1, \ldots, s_n$, $t_1, \ldots, t_n$ are fixed once and for all). In particular, it makes invariant sense to say that a specific $P \in \mathrm{Diff}^2(\mathbb{R}^n)$ has "no linear terms". Indeed, this means that the e_i's and f_i's all equal 0 . Clearly, the a_{ij} , b_{ij} , c_{ij} , d_{ij} , and g may vary with the choice of representation of type (2.8) , but even then, once c_{ij} and d_{ij} are given, g is uniquely determined for P . In fact, g depends on the "diagonal" terms c_{ii} and d_{ii} . Furthermore, it is clear that any $P \in \mathrm{Diff}^2(\mathbb{R}^n)$ has a unique "<u>symmetric representation</u>" of type (2.8), i.e., a representation in which $c_{ij} = d_{ij}$ and

in which $a_{ij} = a_{ji}$ and $b_{ij} = b_{ji}$; for example, we may speak of <u>the</u> constant term g in <u>the</u> symmetric representation of $P \in \mathrm{Diff}^2(\mathbb{R}^n)$. Also it is apparent that the symmetric representation of $P_1 + P_2$ is the "sum" of the symmetric representations of P_1 and P_2 , and similarly for the symmetric representation of cP . Thus, we can make the following definition:

<u>Definition 2.12.</u> $\mathrm{Diff}^2_o(\mathbb{R}^n)$ is the subspace of those $P \in \mathrm{Diff}^2(\mathbb{R}^n)$ having no linear terms and having no constant term in their symmetric representation.

<u>2.</u> We also need to define the "symbol" of $P \in \mathrm{Diff}^2(\mathbb{R}^n)$.

<u>Definition 2.13.</u> The "<u>symbol</u>" of $P \in \mathrm{Diff}^2(\mathbb{R}^n)$ is the polynomial $p \in S^2\mathbb{R}^{2n}$ gotten by replacing D_{s_i} by t_i and then taking the homogeneous of degree two part of the resulting polynomial. It is easy to see that this defines "symbol" unambiguously, independently of the choice of representation of type (2.8).

The group G of symplectic transformations on $\mathbb{R}^{2n}$ acts on $S^2\mathbb{R}^{2n}$ via $p \longmapsto p \circ g^{-1}$, $p \in S^2\mathbb{R}^{2n}$, $g \in G$. We will denote this action by $p \longmapsto gp$. Our aim is to prove

Theorem 2.14. There exists a linear **isomorphism** $\Theta : S^2\mathbb{R}^{2n} \longrightarrow \mathrm{Diff}_o^2\, \mathbb{R}^n$ such that

(2.9) For every $p \in S^2\mathbb{R}^{2n}$, p is the "symbol" of $\Theta(p)$.

(2.10) For every $g \in G$ there exists a unitary map $U_g : L^2(\mathbb{R}^n) \longrightarrow L^2(\mathbb{R}^n)$ such that for every $p \in S^2\mathbb{R}^{2n}$ and for every $g \in G$

$$\Theta\,(gp) = U_g\, \Theta\,(p)\, U_g^{-1}$$

Remarks: 1. Here we are viewing the elements $P \in \mathrm{Diff}_o^2\,(\mathbb{R}^n)$ as densely defined Hilbert space maps on $L^2(\mathbb{R}^n)$, for example, going from $\mathscr{S}$ to $\mathscr{S}$, where $\mathscr{S}$ is the Schwartz space. We shall show that U_g can be chosen in such a way that both U_g and U_g^{-1} map $\mathscr{S}$ into $\mathscr{S}$.

2. We can show that the Ug's can be chosen so that $U_{g_1g_2} = c(g_1,g_2)\, U_{g_1} U_{g_2}$, where $c(g_1,g_2)$ is a complex number of modulus one. Hence, $P \longmapsto g\,P \equiv U_g\, P\, U_g^{-1}$ defines an action of G on $\mathrm{Diff}_o^2\, \mathbb{R}^n$, i.e., $(g_1\, g_2)\, P = g_1(g_2P)$. Then condition (2.10) simply states that Θ is an equivariant map with respect to G . The proof of Theorem 2.14 will proceed via a sequence of lemmas.

Lemma 2.15. For any $p \in S^2\mathbb{R}^{2n}$ there is a unique $P \in \mathrm{Diff}_o^2\, \mathbb{R}^n$ having p as its "symbol" .

Pf.

Clearly, every $p \in S^2\mathbb{R}^{2n}$ has a unique symmetric representation.

$$(2.11) \qquad p = \Sigma a_{ij} t_i t_j + \Sigma b_{ij} s_i s_j + \Sigma c_{ij} s_i t_j + \Sigma c_{ij} t_j s_i$$

[Here (a_{ij}), (b_{ij}) are symmetric matrices]

Then, replace t_j by D_{s_j}, and take

$$(2.12) \qquad P = \Sigma a_{ij} D_{s_i} D_{s_j} + \Sigma b_{ij} s_i s_j + \Sigma c_{ij} s_i D_{s_j} + \Sigma c_{ij} D_{s_j} s_i .$$

Since (a_{ij}), (b_{ij}) are symmetric, and since the coefficients of $s_i D_{s_j}$ and of $D_{s_j} s_i$ are equal, it is clear that $P \in \mathrm{Diff}_o(\mathbb{R}^n)$. Moreover, it is clear that p is the "symbol" of P. This proves existence.

Suppose next that $P \in \mathrm{Diff}_o^2 \mathbb{R}^n$ has $p \in S^2 \mathbb{R}^{2n}$ for its symbol. Then, clearly, if P has the symmetric representation (2.12) then p has the symmetric representation (2.11). But a given $p \in S^2 \mathbb{R}^{2n}$ has a unique symmetric representation (2.11). Thus, $P \in \mathrm{Diff}_o^2(\mathbb{R}^n)$ is uniquely determined by p.

QED

Using Lemma 2.15 we can define a linear map $\Theta : S^2 \mathbb{R}^{2n} \longrightarrow \mathrm{Diff}^2 \mathbb{R}^n_o$ as follows:

<u>Definition 2.16.</u> $\Theta(p)$ is the unique element in $\mathrm{Diff}^2_o \mathbb{R}^n$ having p as its "symbol" .

Thus there is precisely one linear map Θ satisfying condition (2.9) of Theorem 2.14 . We shall see that this Θ also satisfies condition (2.10) . To prove this we need some further lemmas.

Let $S^1 \mathbb{R}^{2n}$ be the polynomials on $\mathbb{R}^{2n}$ homogeneous of degree 1 , i.e., all linear functionals $\ell(\underline{s},\underline{t}) = \Sigma q_i s_i + \Sigma b_i t_i$. Of course G acts on $S^1 \mathbb{R}^{2n}$ via $g\ell \equiv \ell \circ g^{-1}$. Let $\mathrm{Diff}^1_o \mathbb{R}^n$ be the vector space of all linear differential operators on $\mathbb{R}^n$ of the form $\Sigma a_i s_i + \Sigma b_i D_{s_i}$, where a_i, b_i are constants. Clearly $L \in \mathrm{Diff}^1_o \mathbb{R}^n$ uniquely determines and is uniquely determined by its coefficients a_i, b_i . Thus Q given by

$$(2.13) \qquad Q : S^1 \mathbb{R}^{2n} \longrightarrow \mathrm{Diff}^1_o \mathbb{R}^n$$

$$s_i \longmapsto s_i$$

$$t_i \longmapsto D_{s_i}$$

defines a vector space isomorphism. (We could say that for every $\ell \in S^1 \mathbb{R}^{2n}$, $Q(\ell)$ is the unique $L \in \mathrm{Diff}^1_o \mathbb{R}^n$ having

ℓ as its "symbol", "symbol" being defined in an analoguous sense to Def. 2.13). Then we get an action of G on $\mathrm{Diff}^1_0 \mathbb{R}^n$ by transferring the action of G on $S^1 \mathbb{R}^{2n}$. More precisely, for any $g \in G$ **we define the right hand** arrow g in diagram (2.14) below as the unique map which makes the diagram commute.

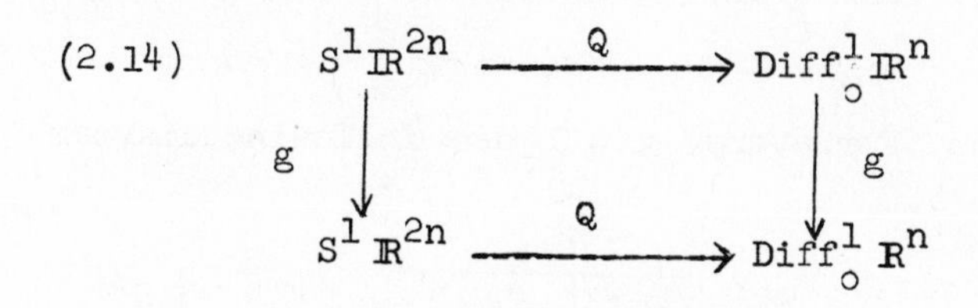

Rather than <u>define</u> a G-action on $\mathrm{Diff}^2_0 \mathbb{R}^n$ by a transfer of the G-action on $S^2 \mathbb{R}^{2n}$, we give a definition in terms of the G-action on $\mathrm{Diff}^1_0 \mathbb{R}^n$ and subsequently show that the action thus defined is the transfer of the G-action on $S^2 \mathbb{R}^{2n}$. Indeed, a key reason for choosing to work with the symmetric representation (and recall that the class $\mathrm{Diff}^2_0 \mathbb{R}^n$ is <u>defined</u> by means of the symmetric representation) is that the G-action on $\mathrm{Diff}^2_0 \mathbb{R}^n$ obtained by transfer of the G-action on $S^2 \mathbb{R}^{2n}$ actually has the expression (2.15), given below, in terms of the G-action on $\mathrm{Diff}^1_0 \mathbb{R}^n$.

<u>Definition 2.17</u> For any $g \in G$ and $P \in \mathrm{Diff}^2_0 (\mathbb{R}^n)$ define $gP \in \mathrm{Diff}^2 (\mathbb{R}^n)$ as follows: If the unique symmetric representation of P is given by (2.12), then

$$(2.15)\qquad gP = \Sigma\, a_{ij}\,(gD_{s_i})(gD_{s_j}) + \Sigma\, b_{ij}\,(gs_i)(gs_j) + \Sigma\, c_{ij}\,(gs_i)(gD_{s_j}) + \Sigma\, c_{ij}\,(gD_{s_j})(gs_i)$$

Here the action of g on the right hand side is that defined by diagram (2.14).

<u>Lemma 2.18.</u> (a) For every $g \in G$ and for every $P \in \mathrm{Diff}_o^2\,\mathbb{R}^n$ $gP \in \mathrm{Diff}_o^2\,\mathbb{R}^n$.

(b) For every $g \in G$ the following diagram commutes:

(2.16)

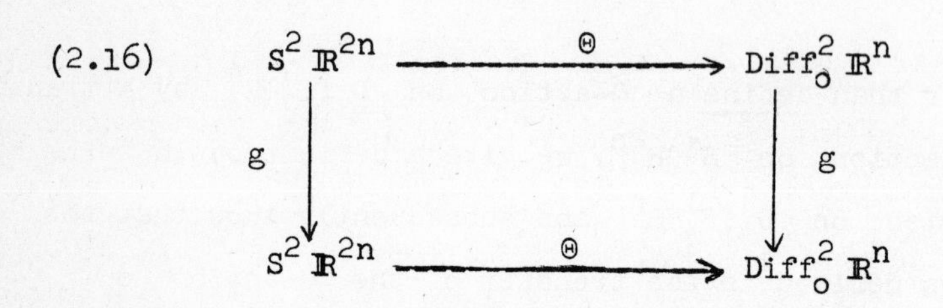

Notice that (b) says that the G-action (2.15) is the transfer of the G-action on $S^2\mathbb{R}^{2n}$.

<u>Pf:</u>

(a) It is clear that gP is well-defined as an element of $\mathrm{Diff}^2\,\mathbb{R}^n$. To see that gP is in fact in $\mathrm{Diff}_o^2\,\mathbb{R}^n$ first note that is is easy to verify that for every $L_1, L_2 \in \mathrm{Diff}_o^1\,\mathbb{R}^n$ $L_1L_2 + L_2L_1$ is an element of $\mathrm{Diff}_o^2\,\mathbb{R}^n$. Second, observe that since (a_{ij}) and (b_{ij}) are symmetric, and since the coefficients of $(gs_i)(gD_{s_j})$ and of $(gD_{s_j})(gs_i)$ are equal, then gP as defined above is a linear combination of terms of the form $L_1L_2 + L_2L_1$.

(b) Let p be an element of $S^2\mathbb{R}^{2n}$. Since $\Theta(gp)$ has gp as its "symbol", and since there is a unique element of $\mathrm{Diff}_o^2\mathbb{R}^n$ with this property, it suffices to prove that $g(\Theta(p))$ has gp as its "symbol". But since diagram (2.14) commutes, gL has $g\ell$ for its "symbol" if L has ℓ for its "symbol", $\ell \in S^1\mathbb{R}^{2n}$, $L \in \mathrm{Diff}_o^1\mathbb{R}^n$. Moreover, let $S^\infty\mathbb{R}^{2n}$ denote the ring of all differential operators on $\mathbb{R}^n$ with polynomial coefficients. Then it is easy to see, upon making the obvious definition of "symbol", that the "symbol" map is a homomorphism from $\mathrm{Diff}^\infty\mathbb{R}^n$ to $S^\infty\mathbb{R}^{2n}$. Furthermore, $g : S^\infty\mathbb{R}^{2n} \longrightarrow S^\infty\mathbb{R}^{2n}$ given by $v \longmapsto v \circ g^{-1}$ is clearly a homomorphism. In view of Def. 2.17 it easily follows that $g(\Theta(p))$ has gp as its "symbol".

QED

We are now ready to prove that $\Theta(gp)$ and $\Theta(p)$ are unitarily equivalent for any $g \in G$ and $p \in S^2\mathbb{R}^{2n}$. In view of Lemma 2.18 it suffices to show that $\Theta(p)$ and $g\,\Theta(p)$ are unitarily equivalent. This will follow, essentially, from the Stone-von Neumann theorem (see, for example, [27]).

Indeed, consider the Heisenberg **algebra** H_n . This is the (nilpotent) Lie algebra over $\mathbb{R}$ with respect to Poisson brackets $\{\ ,\ \}$ generated by $1, s_1, \ldots, s_n, t_1, \ldots, t_n$. Let $\mathscr{H}_n$ be the simply connected Lie group having H_n as its Lie algebra. According to the Stone-von Neumann Theorem for each $\lambda \in \mathbb{R}$, $\lambda \neq 0$, there exists an irreducible unitary representation T_λ of $\mathscr{H}_n$ on $L^2(\mathbb{R}^n)$ such that the corresponding skew-adjoint representation (still denoted T_λ) of H_n takes the form

(2.17) $$T_\lambda : \quad 1 \longmapsto \text{multiplication by } i\lambda$$
$$s_i \longmapsto \text{multiplication by } i\lambda s$$
$$t_i \longmapsto \frac{d}{ds_i}$$

and any irreducible unitary representation T of $\mathscr{H}_n$ (except those trivial on the one-dimensional center of $\mathscr{H}_n$) is unitarily equivalent to T_λ for some λ . (In fact, λ is determined as follows: T being irreducible, the corresponding representation of H_n maps 1 , which is a basis for the center of H_n , to $i\lambda$ for some real $\lambda \neq 0$.) Moreover, any two unitary operators implementing the equivalence between T and T_λ agree up to multiplication by a complex number of absolute value 1 . Now since $g \in G$ is a linear <u>canonical</u> transformation, $g : S^1\mathbb{R}^{2n} \longrightarrow S^1\mathbb{R}^{2n}$ defines a Lie algebra isomorphism

(2.18) $$g : H_n \longrightarrow H_n \quad .$$

Since $\mathcal{N}_n$ is a simply connected nilpotent Lie group the exponential map $\exp : H_n \longrightarrow \mathcal{N}_n$ is bijective (see, for example, [17]). Thus we can define a bijection

$$(2.19) \qquad g : \mathcal{N}_n \longrightarrow \mathcal{N}_n , \quad g : \exp x \longmapsto \exp (gx) , \quad \text{for } x \in H_n .$$

But one can verify that

$$(2.20) \qquad \exp x \exp y = \exp (x + y + 1/2 [x,y]) \text{ for every } x,y \in H_n .$$

Thus the map g of (2.19) is an automorphism of $\mathcal{N}_n$, and so

$$(2.21) \qquad T_1 \circ g : \mathcal{N}_n \longrightarrow \{\text{unitary operators on } L^2 (\mathbb{R}^n)\}$$

defines a unitary representation of $\mathcal{N}_n$. $T_1 \circ g$ is irreducible, for T_1 is irreducible, and T_1 and $T_1 \circ g$ have the same range. Also, since g maps 1 to 1, it follows easily that the representation of H_n corresponding to $T_1 \circ g$ maps 1 to i. (Indeed, the representation of H_n corresponding to $T_1 \circ g$ is the composition with the map g of (2.18) of the representation of H_n corresponding to T_1). Thus, by the Stone-von Neumann theorem, there exists a unitary map $U_g : L^2 (\mathbb{R}^n) \longrightarrow L^2 (\mathbb{R}^n)$ such that

$$(2.22a) \qquad T_1 \circ g = U_g \, T_1 \, U_g^{-1} .$$

The corresponding equality holds for the associated Lie algebra representations. Thus, from the definition (2.13) of Q it follows immediately that

$$(2.22b) \qquad Q(g\ell) = U_g \, Q(\ell) \, U_g^{-1} \quad \text{for every} \quad \ell \in S^1 \mathbb{R}^{2n} .$$

That is, since the diagram (2.14) commutes,

$$(2.23) \qquad g\,(Q(\ell)) = U_g \,(Q(\ell))\, U_g^{-1} \quad \text{for every} \quad \ell \in S^1 \mathbb{R}^{2n} .$$

Using (2.23) we shall prove

$$(2.24) \qquad g\,(\Theta(p)) = U_g \,(\Theta(p))\, U_g^{-1} \quad \text{for every} \quad p \in S^2 \mathbb{R}^{2n} .$$

Indeed, let p have the symmetric representation (2.11). Then we know that $\Theta(p)$ has the symmetric representation

$$\Theta(p) = \Sigma\, a_{ij} D_{s_i} D_{s_j} + \Sigma\, b_{ij} s_i s_j + \Sigma\, c_{ij} s_i D_{s_j} + \Sigma c_{ij} D_{s_j} s_i .$$

Hence, by Def. 2.17, $g\,(\Theta(p))$ has the symmetric representation

$$g(\Theta(p)) = \Sigma a_{ij}\, g(D_{s_i})\, g(D_{s_j}) + \Sigma b_{ij}\, (g\, s_i)(g\, s_j)$$

$$+ \Sigma c_{ij}\, (g\, s_i)(g\, D_{s_j}) + \Sigma c_{ij}\, (g\, D_{s_j})(g\, s_i)$$

$$= (\text{by } 2.23)$$

$$\Sigma a_{ij}\, (U_g\, D_{s_i}\, U_g^{-1})(U_g\, D_{s_j}\, U_g^{-1}) + \dots$$

$$= U_g\, (\Sigma a_{ij}\, D_{s_i}\, D_{s_j} + \dots)\, U_g^{-1}$$

$$= U_g\, \Theta(p)\, U_g^{-1} \quad .$$

This proves (2.24) and thus concludes the proof of Theorem 2.14 .

QED

Remarks: 1. The uniqueness of U_g in (2.23) up to multiplication by a complex number of modulus 1 yields the fact that, no matter how we choose the U_g's , we get that $U_{g_1 g_2} = c(g_1, g_2)\, U_{g_1}\, U_{g_2}$ where $c(g_1, g_2)$ is a complex number of modulus 1 . Hence, as stated in Remark 2. following Theorem 2.14 , $g : P \longmapsto U_g\, P\, U_g^{-1}$ actually defines an action of G on $\mathrm{Diff}_o^2\, \mathbb{R}^n$, and $\Theta : S^2\, \mathbb{R}^{2n} \longrightarrow \mathrm{Diff}_o^2\, \mathbb{R}^n$ is a G-equivariant map. Since our original construction of the test-operators, we have learned that the use of the

Stone-von Neumann theorem to prove (2.22a) is classical, and that, in fact, by work of Weil and Shale, the projective representation U_g of the symplectic group G may be lifted to a genuine (i.e., with $c \equiv 1$) unitary representation of a two-fold covering group, the metaplectic group. (Actually, in our original proof of (2.22a) we made use of an incorrect version of the Stone-von Neumann theorem, stated in terms of the Lie algebra H_n rather than the Lie group $\mathcal{H}_n$, and so we worked directly with (2.18) .)

2. If follows from (2.22a) that U_g and U_g^{-1} map the Schwartz space $\mathcal{S}$ into itself. In fact, as pointed out in [18], equals the space of C^∞ vectors in $L^2(\mathbb{R}^n)$ with respect to the representation T_1, i.e., the space of all $f \in L^2(\mathbb{R}^n)$ such that the map $a \longmapsto T_1(a)f$ from $\mathcal{H}_n$ to $L^2(\mathbb{R}^n)$ is C^∞ . This characterization of $\mathcal{S}$, together with (2.22a) easily yields the desired result.

3. An alternate proof of (2.23) not using the Stone-von Neumann theorem, exactly along the lines of the proof in [23] of the unitary equivalence of test-complexes, explicitly exhibits U_g as an integral operator. In particular, one can see directly that U_g and U_g^{-1} map $\mathcal{S}$ into itself.

4. It is easy to see, if we define $S^q\,\mathbb{R}^{2n}$ and $\mathrm{Diff}^q\,\mathbb{R}^n$

is the obvious way, that each $P \in \mathrm{Diff}^q \mathbb{R}^n$ has a unique "symmetric representation" . If we then define $\mathrm{Diff}^q_o \mathbb{R}^n$ as the subspace of those $P \in \mathrm{Diff}^q \mathbb{R}^n$ having no terms of "order" less than q in their unique "symmetric representation", then by following the same procedure as in the proof of Theorem 2.14 we can prove the same theorem, but with q replacing 2 .

Returning to our specific question of interest, we see from Theorem 2.14 how to associate invariantly, up to unitary equivalence, a Hilbert space operator $\tilde{P}_{(x,\xi)}$ in $\mathrm{Diff}^2_o \mathbb{R}^n$ to $\tilde{p}_{(x,\xi)}$ for each $(x,\xi) \in \Sigma$. We need to add on to $\tilde{P}_{(x,\xi)}$ an invariantly defined constant term. We shall get this term, essentially, by taking the sub-principal part of P , our original operator, evaluated at (x,ξ) . Recall (see, for example [6]) that if we have a differential operator P mapping sections of the 1/2-density bundle $\Omega_{1/2}$ on X to sections of the 1/2-density bundle $\Omega_{1/2}$ on X then there is an invariantly defined (independent of choice of local coordinates on X) sub-principal part of P which is a function from $T^*X\backslash 0$ into $\mathrm{Hom}\,(\Omega_{1/2},\Omega_{1/2})$ given, in any choice of local coordinates $x_1,\ldots\ldots,x_k$ on X (with the canonically corresponding cotangent variables $\xi_1,\ldots\ldots,\xi_k$, and with the canonically corresponding frame for $\Omega_{1/2}$) by

$$(2.25) \qquad \sigma_{sub}(P) = \sigma_{m-1}(P) - \frac{1}{2\sqrt{-1}} \sum_{\ell} \frac{\partial^2}{\partial \xi_\ell \partial x_\ell} \sigma_m(P)$$

where σ_m denotes the top symbol and σ_{m-1} the (m-1) - th order symbol of P, with respect to the given coordinates and frame.

Since our operator P goes from functions to functions and not from 1/2-densities to 1/2-densities, we do not have an invariantly defined subprincipal part. We shall see, nevertheless, that $\sigma_{sub}(P)|_{\Sigma}$, the restriction of $\sigma_{sub}(P)$ to Σ is invariantly defined since p vanishes to second order on Σ. [All that is needed is that p vanishes to at least second order.] Before proceeding to show this we first digress and discuss a partial counterpart at the germ level of the essentially first-order jet level result, Theorem 2.14. (We shall discuss the "general" version with q replacing 2, indicated in Remark 4. above.)

To begin, we observe that the ambiguity in defining $\sigma_{sub}(P)$ when P goes between sections of the trivial bundle rather than between sections of $\Omega_{1/2}$ is exactly analoguous to the ambiguity in defining $\tilde{P}_{(x,\xi)}$, i.e., $\tilde{P}_{(x,\xi)}$ is given only up to unitary equivalence. This is already built into the structure of the 1/2-density bundle since, corresponding to each choice of local coordinates in X there is a canonical choice of frame for $\Omega_{1/2}$. This leads us to the analogue of our G-equivariant map $\Theta : S^q\mathbb{R}^{2n} \longrightarrow \mathrm{Diff}^q_o\mathbb{R}^n$. In stead of the group G of linear canonical transformations on $\mathbb{R}^{2n}$ (with fixed coordinates $s_1,\dots,s_n$, $t_1,\dots t_n$) we consider the pseudogroup $\mathcal{S}$ of local diffeomorphisms on $\mathbb{R}^n$ (with fixed coordinates $x_1,\dots,x_n$). We kepp $L^2(\mathbb{R}^n)$. Instead of Diff_0^q we consider the space, $\mathcal{D}\mathrm{iff}_0^q\,\mathbb{R}^n$. $\mathcal{D}\mathrm{iff}_0^q\,\mathbb{R}^n$ is the space of qth-order differential operators P from the trivial bundle to the trivial bundle having 0 subprincipal part, where by subprincipal part of P we mean $\sigma_{q-1}(P) - \frac{1}{2\sqrt{-1}} \sum_{\ell} \frac{\partial^2}{\partial\xi_\ell\,\partial x_\ell} \sigma_q(P)$ with respect to the fixed coordinates $x_1,\ldots,x_n$ on $\mathbb{R}^n$ and corresponding cotangent variables $\xi_1,\ldots,\xi_n$. Instead of $S^q\,\mathbb{R}^{2n}$ we consider the sections (denoted by the underlining) $\underline{S^q(T^*\mathbb{R}^n\backslash 0)}$ of $S^q(T^*\mathbb{R}^n\backslash 0)$, the space of symbols positive homogeneous of degree q in $\xi_1,\ldots,\xi_n$. (We consider fixed once and for all the coordinates $x_1,\ldots x_n$, $\xi_1,\ldots\xi_n$ on $T^*\mathbb{R}^n\backslash 0$.) We get an action of $\mathcal{G}$ on $\underline{S^q(T^*\mathbb{R}^n\backslash 0)}$ by

$$(2.26) \qquad g : p \longmapsto p \circ \tilde{g}^{-1}$$

where $\tilde{g}$ is the local diffeomorphism of $T^*\mathbb{R}^n\backslash 0$ induced by the local diffeomorphism g of $\mathbb{R}^n$. We get a representation of $\mathcal{G}$ as unitary operators on $L^2(\mathbb{R}^n)$ by

$$(2.27) \qquad U_g : f \longmapsto |J_g{-1}|^{1/2}\,(f \circ g^{-1})$$

where $J_g{-1}$ denotes the Jacobian determinant of the local diffeomorphism g^{-1} of $\mathbb{R}^n$. Notice that multiplication by $|J_g{-1}|^{1/2}$ corresponds to changing to the canonical frame for $\Omega_{1/2}$ associated to the new coordinates given by the local diffeomorphism g . We next define an action of $\mathcal{G}$

on $\mathcal{D}iff_0^q \mathbb{R}^n$ by

$$(2.28) \qquad g : P \longmapsto gP = U_g \, P \, U_g^{-1}$$

That the $\mathcal{G}$-action thus defined actually maps $\mathrm{Diff}_0^q \mathbb{R}^n$ into itself follows directly from the fact that "sub-principal part" is an invariant notion for differential operators between sections of $\Omega_{1/2}$.

Let $\mathcal{D}iff^{q-2} \mathbb{R}^n$ be the differential operators of order $q-2$ on $\mathbb{R}^n$ from the trivial bundle to itself. $\mathcal{D}iff^{q-2} \mathbb{R}^n$ is contained in $\mathrm{Diff}_0^q \mathbb{R}^n$, and the $\mathcal{G}$-action (2.28) takes $\mathrm{Diff}^{q-2} \mathbb{R}^n$ into itself. The analogue of the G-equivariant map Θ of Theorem 2.14 is the $\mathcal{G}$-equivariant map.

$$(2.29) \qquad \psi : \underline{S^q \, (T^* \mathbb{R}^n \backslash 0)} \longrightarrow \mathcal{D}iff_0^q \mathbb{R}^n \, / \, \mathcal{D}iff^{q-2} \mathbb{R}^n$$

where $\psi(p)$ is defined as the unique equivalence class containing an element of $\mathrm{Diff}_0^q \mathbb{R}^n$ which has p as its principal symbol.

Of course, the preceding is only a partial analogue to Theorem 2.14. since we have made a fixed distinction between space variables $x_1, \ldots, x_n$ and cotangent variables $\xi_1, \ldots, \xi_n$, and we have considered only those local diffeomorphisms of $T^* \mathbb{R}^n \backslash 0$ induced by local diffeomorphisms of $\mathbb{R}^n$. The full analogue of Theorem 2.14 would replace G by the pseudogroup

of all local canonical transformations of $T^*\mathbb{R}^n\backslash 0$.
Perhaps the work of Kostant and Ste nberg (see, for example, [18]) will be relevant here.

We now return to the problem of defining an invariant constant term to add on to $\tilde{P}_{(x,\xi)}$.

Proposition 2.19. Let P be of type (2.1) and suppose that p vanishes to at least second order on Σ. Then the restriction to Σ of the "sub-principal part" of P

$$(2.30) \qquad \sigma_{sub}(P)|_\Sigma \equiv \sigma_{m-1}(P) - \frac{1}{2\sqrt{-1}} \sum_\ell \frac{\partial^2}{\partial\xi_\ell \partial x_\ell} \sigma_m(P)|_\Sigma$$

is an intrinsically defined invariant associated to P, independent of the choice of local coordinates $x_1,\ldots,x_k$ on X, even though P goes between sections of the trivial bundle of X rather than between sections of the 1/2-density bundle on X.

Pf:

The proof will consist of showing that $\sigma_{sub}(P)|_\Sigma = \sigma_{m-1}(S)|_\Sigma$, where $\sigma_{m-1}(S)|_\Sigma$ is an invariant obtained as follows:

Fix $(x,\xi) \in \Sigma$ and choose local C^∞ - parameters

$u_1,\ldots,u_{2n}$ defining Σ in a conic neighborhood Γ of (x,ξ) in $T^*X\backslash 0$. For convenience we take the u_i's (positive) homogeneous of degree 1/2 in ξ. Then (possibly by passing to a smaller neighborhood Γ) since p vanishes at least to second order on Σ we can find C^∞ function a_{ij}, which clearly we may assume <u>symmetric in i and j</u>, homogeneous of degree $m-1$ such that in Γ

$$(2.31) \qquad p = \Sigma\, a_{ij}\, u_i\, u_j$$

Next choose pseudo-differential operators U_i, $i = 1,\ldots,2n$ of order 1/2 and A_{ij} of order $n-1$ such that $\sigma_{1/2}(U_i) = u_i$ and $\sigma_{m-1}(A_{ij}) = a_{ij}$ in Γ (again possibly shrinking). Then define the pseudo-differential operator S by

$$(2.32) \qquad S = P - \Sigma\, A_{ij}\, U_i\, U_j$$

Since $\sigma_m(S) \equiv 0$ in Γ, $\sigma_{m-1}(S)$ is well-defined on Γ, independent of any choice of local coordinates on X, being, in fact, the <u>principal</u> symbol of S. In fact, the following holds:

<u>Lemma 2.20.</u> $\sigma_{m-1}(S)|_\Sigma$ is an invariant attached to P.

That is, in addition to being independent of choice of local coordinates on X it is independent of the choice of a_{ij} and u_i used in (2.31), assuming however that a_{ij} is symmetric in i and j, and independent of the choice of A_{ij}, U_i having a_{ij} and u_i as principal symbol

Pf:

Suppose we write $S = P - \Sigma A_{ij} U_i U_j$. Choose local coordinates $x_1, \ldots, x_k$ in X. (We already know that $\sigma_{m-1}(S)$ is well-defined independently of any choice of local coordinates on X.) Next, by using the symbol calculus for pseudo-differential operators we get that

$$\sigma_{m-1}(S)|_\Sigma = \sigma_{m-1}(P)|_\Sigma - \sigma_{m-1}(\Sigma A_{ij} U_i U_j)|_\Sigma$$

$$= \sigma_{m-1}(P)|_\Sigma - \Sigma[\sigma_{m-1}(A_{ij})\,\sigma_o(U_i U_j) + \sigma_{m-2}(A_{ij})\,\sigma_1(U_i U_j)$$

$$+ \frac{1}{\sqrt{-1}} \sum_\ell \frac{\partial}{\partial \xi_\ell} \sigma_{m-1}(A_{ij}) \frac{\partial}{\partial x_\ell} \sigma_1(U_i U_j)]|_\Sigma \quad .$$

Since $\sigma_1(U_i U_j) = \sigma_{1/2}(U_i)\,\sigma_{1/2}(U_j) = u_i u_j$, $\sigma_1(U_i U_j)|_\Sigma = 0$

and $\dfrac{\partial}{\partial x_\ell} \sigma_1(U_i U_j)|_\Sigma = 0$.

Thus, $\sigma_{m-1}(S)|_\Sigma = \sigma_{m-1}(P)|_\Sigma - \Sigma\, \sigma_{m-1}(A_{ij})\ \sigma_o(U_i U_j)|_\Sigma$

But $\sigma_{m-1}(A_{ij}) = a_{ij}$ and

$$\sigma_o(U_i U_j)|_\Sigma = \sigma_{1/2}(U_i)\,\sigma_{-1/2}(U_j)|_\Sigma + \sigma_{-1/2}(U_i)\sigma_{1/2}(U_j)|_\Sigma$$

$$+ \frac{1}{\sqrt{-1}} \sum_\ell \frac{\partial}{\partial \xi_\ell} \sigma_{1/2}(U_i) \frac{\partial}{\partial x_\ell} \sigma_{1/2}(U_j)|_\Sigma .$$

Hence, again since $\sigma_{1/2}(U_i)|_\Sigma = u_i|_\Sigma = 0$, we get that

$$\sigma_o(U_i U_j)|_\Sigma = \frac{1}{\sqrt{-1}} \sum_\ell \frac{\partial u_i}{\partial \xi_\ell} \frac{\partial u_j}{\partial x_\ell}\Big|_\Sigma .$$

That is,

$$\sigma_{m-1}(S)|_\Sigma = \sigma_{m-1}(P)|_\Sigma - \frac{1}{\sqrt{-1}} \sum_\ell \sum_{i,j} a_{ij} \frac{\partial u_i}{\partial \xi_\ell} \frac{\partial u_j}{\partial x_\ell}\Big|_\Sigma . \tag{2.33}$$

Thus, in view of the symmetry of a_{ij} in i and j, it is easy to check that

$$\sigma_{m-1}(S)|_\Sigma = \sigma_{sub}(P)|_\Sigma \tag{2.34}$$

Hence, in order to prove both Lemma 2.20 and Prop. 2.19 it suffices to show that if we can write p in a neighborhood of $(x,\xi) \in \Sigma$ both as $p = \sum_{i,j} a_{ij} u_i u_j$ and as $p = \sum_{i,j} \tilde{a}_{ij} \tilde{u}_i \tilde{u}_j$ where a_{ij} and $\tilde{a}_{ij}$ each is symmetric in i and j, and where $u_1,\dots,u_{2n}$ and $\tilde{u}_1,\dots\tilde{u}_{2n}$ each is a family of local parameters defining Σ , then for every ℓ

$$(2.35) \qquad \sum_{i.j} a_{ij} \frac{\partial u_i}{\partial \xi_\ell} \frac{\partial u_j}{\partial x_\ell} \Big|_\Sigma = \sum_{i,j} \tilde{a}_{ij} \frac{\partial \tilde{u}_i}{\partial \xi_\ell} \frac{\partial \tilde{u}_j}{\partial x_\ell} \Big|_\Sigma$$

But since a_{ij} and $\tilde{a}_{ij}$ are each symmetric in i and j we see that for every $(x,\xi) \in \Sigma$ and for every ℓ

$$(2.36) \qquad \sum_{i,j} a_{ij} \frac{\partial u_i}{\partial \xi_\ell} \frac{\partial u_j}{\partial x_\ell} \Big|_{(x,\xi)}$$

$$= 1/2 \text{ Hess } p\Big|_{(x,\xi)} \left(\frac{\partial}{\partial \xi_\ell}, \frac{\partial}{\partial x_\ell}\right)$$

$$= \sum_{i,j} \tilde{a}_{ij} \frac{\partial \tilde{u}_i}{\partial \xi_\ell} \frac{\partial \tilde{u}_j}{\partial x_\ell} \Big|_{(x,\xi)} .$$

<u>QED</u>

<u>Note:</u> S differs from the lower order part Z_o in [4] precisely in the same fashion that $\tilde{p}_{(x,\xi)}$ differs from the symmetric form p_o in [4] .

§2.3 Proof of theorem

In §2.5 we shall prove Theorem 2.4 by computing the eigenvalues of the $\widetilde{P}_{(x,\xi)}$, and thus showing explicitly that the criterion of Theorem 2.4 agrees with the criterion of Boutet de Monvel and Treves ([3], [4]) for hypoellipticity with loss of one derivative. In the present section we shall give a different proof, which combines results of Grushin ([13], [14]) in the "flat" case with the results of [3], [4] in the "non-flat" case. The proof will draw not so much on the specific hypoellipticity criterion of [3], [4] but rather on the determination in [3], [4] of what "ingredients" from P are needed for the hypoellipticity criterion, and on the microlocalizability results of [3], [4]. The passage from the "non-flat" context of [3], [4] to the "flat" context of [13], [14] is accomplished by associating to $\widetilde{P}_{(x,\xi)} + \sigma_{sub}(P)|_{(x,\xi)}$ an operator $\check{P}_{(x,\xi)}$ satisfying the hypotheses of Grushin, in particular the "quasi-homogeneity", and having $\widetilde{P}_{(x,\xi)} + \sigma_{sub}(P)|_{(x,\xi)}$ as its partial Fourier transform. Actually, at one point (see the proof of (2.68)) our argument is not quite complete. It seems clear to us that the details can be carried out fully with some additional effort; however, since the direct verification of Theorem 2.4 in §2.5 is available, we have not attempted to do so. Nevertheless, we have felt it worthwhile to present our argument here, for it illustrates how the seemingly special quasi-homogeneous differential operators with polynomial coefficients of Grushin appear naturally in the study of the "non-flat" case in conjunction with the test-operators. Turning the argument around, we note that the "flat" cases that Grushin treats are much more general than the "flat" version of [3] , [4] . This, together with the relationship of $\tilde{P}_{(x,\xi)} + \sigma_{sub}(P)|_{(x,\xi)}$ to $\check{P}_{(x,\xi)}$, suggests that the quasi-homogeneous differential operators of Grushin may provide a clue for the construction of "test-operators" in more general "non-flat" cases. It appears likely that for these more general cases Theorem 2.4 still holds (with the appropriate form of hypoellipticity substituted for hypoellipticity with loss of 1 derivative) , though, of course, the explicit form and manner of computation of the eigenvalues of the test-operators may be quite different from what occurs in §2.5 . We shall also make use of the results of Grushin in our discussion in §2.4 of how the index of the test-operators $\tilde{P}_{(x,\xi)} + \sigma_{sub}(P)|_{(x,\xi)}$ measures the difference between the hypoellipticity and local solvability properties of P .

We begin by computing what $\tilde{p}_{(x,\xi)}$ and $\omega_{(x,\xi)}$ look like in "local coordinates". Fix $(x_0,\xi_0) \in \Sigma$. Then we can choose local parameters $u_1,\ldots,u_n$, $v_1,\ldots,v_n$ defining Σ in a conic neighborhood of (x_o,ξ_o) in $T^*X\backslash 0$ and satisfying the canonical commutation relations at (x_o,ξ_o)

$$(2.37) \qquad \{u_i,u_j\}|_{(x_o,\xi_o)} = 0 \quad , \quad \{v_i,v_j\}|_{(x_o,\xi_o)} = 0$$

$$\{u_i,v_j\}|_{(x_o,\xi_o)} = -\delta_{ij} \quad .$$

This follows easily from the fact that for any choice of local parameters, satisfying (2.37) or not, $du_1,\ldots,du_n$ $dv_1,\ldots,dv_n$ form local frames for the bundle $N(\Sigma)$, and from the fact that the non-degenerate symplectic form $\omega(x_o,\xi_o)$ on $N(\Sigma)_{(x_o,\xi_o)}$ is given by $\omega_{(x_o,\xi_o)}(df,dg) = \{f,g\}_{(x_o,\xi_o)}$ for df and dg in $N(\Sigma)_{(x_o,\xi_o)}$. (See Remark 2.7 .) In fact, as is shown in [4] , we can actually choose local parameters for Σ so that the canonical commutation relations (2.37) hold in a full conic neighborhood of (x_o,ξ_o) rather than just at the point (x_o,ξ_o) itself.

We use $du_1,\ldots,du_n$, $dv_1,\ldots,dv_n$ as a basis in $N(\Sigma)_{(x_o,\xi_o)}$. Thus any element $v \in N(\Sigma)_{(x_o,\xi_o)}$ may be uniquely expressed as

$$(2.38) \qquad v = \Sigma\, s_i\, du_i + \Sigma\, t_i\, dv_i$$

Let us "compute" $\omega_{(x_o,\xi_o)}$.

$$\omega_{(x_o,\xi_o)} \left(\Sigma s_i \, du_i + \Sigma t_i \, dv_i \, , \, \Sigma s_i' \, du_i + \Sigma t_i' \, dv_i\right)$$

$=$ (in view of (2.37) and Remark 2.7)

$$\Sigma t_i s_i' - \Sigma t_i' s_i$$

Hence, if, as in the discussion following Remark 2.7, we now view the symplectic vector space $N(\Sigma)_{(x_o,\xi_o)}$ as a symplectic manifold and make the canonical identification of a vector space with its tangent space at each point, we get that the symplectic 2-form $\omega_{(x_o,\xi_o)}$ on the manifold $N(\Sigma)_{(x_o,\xi_o)}$ is in terms of the coordinates $s_1,\ldots,s_n$, $t_1,\ldots t_n$ given by

$$\omega_{(x_o,\xi_o)} = \Sigma \, dt_i \wedge ds_i \tag{2.39}$$

Thus, in particular, $s_1,\ldots,s_n$, $t_1,\ldots t_n$ form canonical coordinates for the symplectic manifold $N(\Sigma)_{(x_o,\xi_o)}$.

Since p vanishes to second order on Σ we may, in a conic neighborhood of (x_o,ξ_o), express p as

$$(2.40) \qquad p = \Sigma\, a_{ij}\, v_i\, v_j + \Sigma\, b_{ij}\, u_i\, u_j + \Sigma\, 2c_{ij}\, u_i\, v_j$$

Here, a_{ij}, b_{ij}, c_{ij} are C^∞ functions, and a_{ij} and b_{ij} each is <u>symmetric</u> in i and j. (This corresponds, at the germ level, to the "symmetric representation" used in the construction of our test-operators.) Now, using the canonical commutation relations (2.37) and Remark 2.11 we "compute" $\tilde{p}_{(x_o,\xi_o)}$.

Remembering that $u_1, \ldots. u_n$, $v_1, \ldots. v_n$ all vanish on Σ we get

$$\tilde{p}_{(x_o,\xi_o)}\,(du_{i_o}, du_{j_o}) \equiv 1/2\{u_{i_o}, \{u_{j_o}, p\}\}|_{(x_o,\xi_o)}$$

$$= 1/2\, \{u_{i_o}, \{u_{j_o}, \Sigma a_{ij}\, v_i\, v_j\}\}|_{(x_o,\xi_o)} + 1/2\, \{u_{i_o}, \{u_{j_o},$$

$$\Sigma\, 2\, c_{ij}\, u_i\, v_j\}\}|_{(x_o,\xi_o)}$$

$$= 1/2\, \Sigma a_{ij}\, \{u_{i_o}, \{u_{j_o}, v_i\}\, v_j + \{u_{j_o}, v_j\}\, v_i\}\, |_{(x_o,\xi_o)}$$

$$+ 1/2\, \Sigma\, 2c_{ij}\, \{u_{j_o}, u_i\{u_{j_o}, v_j\}\}|_{(x_o,\xi_o)}$$

= (after simplifying the first term and noting that the second term vanishes)

$$1/2 \quad \Sigma\, a_{i_j} \, [\{u_{j_o}, v_i\}\, \{u_{i_o}, v_j\} + \{u_{j_o}, v_j\}\, \{u_{i_o}, v_i\}]|_{(x_o,\xi_o)}$$

$$= 1/2(a_{j_o i_o} + a_{i_o j_o}) = \text{(since } (a_{ij}) \text{ is symmetric)}$$

$$a_{i_o j_o}|_{(x_o,\xi_o)} \quad .$$

In exactly the same way we see that $\widetilde{p}_{(x_o,\xi_o)}(dv_{i_o}, dv_{j_o}) =$ $b_{i_o j_o}\big|_{(x_o,\xi_o)}$. Finally, we compute $\widetilde{p}_{(x_o,\xi_o)}\,(du_{i_o}, dv_{j_o})$.

$$\widetilde{p}_{(x_o,\xi_o)}\,(du_{i_o} dv_{j_o}) \equiv 1/2 \;\{u_{i_o}, \{v_{j_o}, p\}\}|_{(x_o,\xi_o)}$$

$$= 1/2 \;\{u_{i_o}, \{v_{j_o},\; \Sigma\, b_{ij}\, u_i\, u_j\}\}|_{(x_o,\xi_o)} \; +$$

$$1/2 \;\{u_{i_o}, \{v_{j_o},\; \Sigma\, 2c_{ij}\, u_i\, v_j\}\}|_{(x_o,\xi_o)}$$

$$= 1/2 \quad \Sigma\, b_{ij}\; \{u_{i_o}, \{v_{j_o}, u_i\}\, u_j + \{v_{j_o}, u_j\} u_i\}|_{(x_o,\xi_o)}$$

$$+1/2 \quad \Sigma\, 2c_{ij}\; \{u_{i_o}, \{v_{j_o}, u_i\}\, v_j\}|_{(x_o,\xi_o)}$$

= (after simplifying the second term and noting that the first term vanishes)

$$1/2 \;\Sigma\, 2c_{ij}\; [\{v_{j_o}, u_i\}\; \{u_{i_o}, v_j\}]|_{(x_o,\xi_o)}$$

$$= -c_{j_o i_o}|_{(x_o,\xi_o)} \quad .$$

We can now compute the polynomial $\tilde{p}_{(x_o,\xi_o)}(v) \equiv \tilde{p}_{(x_o,\xi_o)}(v,v)$ in terms of the coordinates $s_1,\ldots,s_n$, $t_1,\ldots,t_n$ on $N(\Sigma)_{(x_o,\xi_o)}$. We get

$$(2.41) \qquad \tilde{p}_{(x_o,\xi_o)}(\underline{s},\underline{t}) \equiv \tilde{p}_{(x_o,\xi_o)}(\Sigma s_i\, du_i + \Sigma t_i\, dv_i)$$

$$= \sum_{i,j} \tilde{p}_{(x_o,\xi_o)}(dv_i, dv_j)\, t_i\, t_j + \sum_{i,j} \tilde{p}_{(x_o,\xi_o)}(du_i, du_j)\, s_i s_j$$

$$+ \sum_{i,j} 2\tilde{p}_{(x_o,\xi_o)}(du_i, dv_j)\, s_i\, t_j$$

= (by the preceding computation)

$$\Sigma\, b_{ij}|_{(x_o,\xi_o)}\, t_i\, t_j + \Sigma\, a_{ij}|_{(x_o,\xi_o)}\, s_i\, s_j -$$

$$\Sigma\, 2\, c_{ji}|_{(x_o,\xi_o)}\, s_i\, t_j$$

Since p has the expression (2.40) in terms of the local parameters $u_1,\ldots,u_n$, $v_1,\ldots,v_n$ for Σ satisfying (2.37) we might have hoped that $\tilde{p}_{(x_o,\xi_o)}$ with respect to the canonical coordinates $s_1,\ldots,s_n$, $t_1,\ldots,t_n$ on $N(\Sigma)_{(x_o,\xi_o)}$ would have the form

$$\tilde{p}_{(x_o,\xi_o)}(\underline{s}, \underline{t}) = \Sigma\, a_{ij}|_{(x_o,\xi_o)}\, t_i\, t_j + \Sigma\, b_{ij}|_{(x_o,\xi_o)}\, s_i\, s_j +$$

$$\Sigma\, 2\, c_{ij}|_{(x_o,\xi_o)}\, s_i\, t_j$$

rather than the "twisted" form that it appears to have in (2.41) .

But, in the notation of Theorem 2.14 , for any symplectic transformation g , $\Theta(g\,\tilde{p}_{(x_0,\xi_0)})$ is a representative of the unitary equivalence class of operators $\tilde{P}_{(x_0,\xi_0)}$.

Taking g as the symplectic transformation $(\underline{s},\underline{t}) \longmapsto (\underline{t},-\underline{s})$ we see in view of (2.41) that

$$(2.42)\quad g\,\tilde{p}_{(x_0,\xi_0)}(\underline{s},\underline{t}) \equiv (\tilde{p}_{(x_0,\xi_0)} \circ g^{-1})(\underline{s},\underline{t}) =$$

$$\Sigma\, a_{ij}|_{(x_0,\xi_0)}\, t_i\, t_j + \Sigma\, b_{ij}|_{(x_0,\xi_0)}\, s_i\, s_j +$$

$$\Sigma\, 2\, c_{ij}|_{(x_0,\xi_0)}\, s_i\, t_j \quad .$$

Hence, $\Theta(g\,\tilde{p}_{(x_0,\xi_0)})$, which by abuse of notation we denote by $\tilde{P}_{(x_0,\xi_0)}$, is given by

$$(2.43)\quad \tilde{P}_{(x_0,\xi_0)} = \Sigma\, a_{ij}|_{(x_0,\xi_0)}\, D_{s_i} D_{s_j} + \Sigma\, b_{ij}|_{(x_0,\xi_0)}\, s_i\, s_j$$

$$+ \Sigma\, c_{ij}|_{(x_0,\xi_0)}\, s_i\, D_{s_j} + \Sigma\, c_{ij}|_{(x_0,\xi_0)}\, D_{s_j}\, s_i$$

We next adjoin an additional variable r to the variables $s_1,\ldots,s_n$, and construct the auxiliary differential operator

$$(2.44)\quad \check{P}_{(x_o,\xi_o)}(\underline{s}, D_{\underline{s}}, D_r) \equiv$$

$$\Sigma\, a_{ij}|_{(x_o,\xi_o)} D_{s_i} D_{s_j} + \Sigma\, b_{ij}|_{(x_o,\xi_o)} s_i s_j D_r^2$$

$$+ \Sigma\, c_{ij}|_{(x_o,\xi_o)} s_i D_{s_j} D_r + \Sigma\, c_{ij}|_{(x_o,\xi_o)} D_{s_j} s_i D_r$$

$$+ \sigma_{sub}(P)|_{(x_o,\xi_o)} D_r$$

having as its partial Fourier transform with respect to r essentially the test-operator $\tilde{P}_{(x_o,\xi_o)} + \sigma_{sub}(P)|_{(x_o,\xi_o)}$. (Strictly speaking, we have associated $\check{P}_{(x_o,\xi_o)}$ not to the test-operator, which is a unitary equivalence class, but to a specific representative of that equivalence class. There may be some point to trying to get an invariant object by examining how $\check{P}_{(x_o,\xi_o)}$ varies with the choice of representative for the test-operator. However, we shall not need this refinement for our present application.)

We next recall some of the results of Grushin ([13] and [14]) on local solvability and hypoellipticity for "quasi-homogeneous" operators. We shall not state them in the most general form but only in the generality which we shall need to draw upon.

Grushin considers operators of the form

$$(2.45)\quad L(\underline{y}, D_{\underline{x}}, D_{\underline{y}}) = \sum_{\substack{|\alpha|+|\beta| \leq m \\ |\gamma| \leq m\delta}} a_{\alpha\beta\gamma}\, \underline{y}^{\gamma} D_{\underline{x}}^{\beta} D_{\underline{y}}^{\alpha}$$

where m is an integer, $\delta > 0$, δm is an integer, and $a_{\alpha\beta\gamma}$ is a constant. Let $\underline{\xi}$ denote the dual variable to $\underline{x}$ and $\underline{\eta}$ the dual variables to $\underline{y}$. Then Grushin assumes the total symbol $L(\underline{y}, \underline{\xi}, \underline{\eta})$ of $L(\underline{y}, D_{\underline{x}}, D_{\underline{y}})$ is <u>quasi-homogeneous</u>, i.e.

$$(2.46)\quad L(\underline{y}/\lambda, \lambda^{1+\delta}\, \underline{\xi}, \lambda\underline{\eta}) = \lambda^{m} L(\underline{y}, \underline{\xi}, \underline{\eta}) \text{ for every } \lambda > 0 .$$

He also insists on the following condition, which he refers to as <u>Condition 1</u>:

$$(2.47)\quad L(\underline{y}, D_{\underline{x}}, D_{\underline{y}}) \text{ is elliptic for } \underline{y} \neq 0 .$$

In view of (2.46) the principal symbol $L^{\circ}(\underline{y}, \underline{\xi}, \underline{\eta})$ of L is given by

$$L^{\circ}(\underline{y}, \underline{\xi}, \underline{n}) = \sum_{|\alpha|+|\beta| = m}\ \sum_{|\gamma| = \delta|\beta|} a_{\alpha\beta\gamma}\, \underline{y}^{\gamma}\, \underline{\xi}^{\beta}\, \underline{\eta}^{\alpha} ,$$

so we can express (2.47) as

$$(2.47a) \quad \sum_{|\alpha|+|\beta| = m} \; \sum_{|\gamma| = \delta|\beta|} a_{\alpha\beta\gamma} \, \underline{y}^{\gamma} \, \underline{\xi}^{\beta} \, \underline{\eta}^{\alpha} \neq 0 \qquad \text{for } y \neq 0, \; |\xi|+|\eta| \neq 0$$

Let $L(\underline{y}, \underline{\xi}, D_{\underline{y}})$ denote the partial Fourier transform of $L(\underline{y}, D_{\underline{x}}, D_{\underline{y}})$ with respect to the variables $\underline{x}$. Also, let $H_{(m,\delta)}(R^n_{\underline{y}})$ denote the "weighted Sobolev space" on $R^n_{\underline{y}}$ consisting of all distributions $v(\underline{y})$ satisfying

$$(2.48) \quad (1 + |\underline{y}|)^{(m-|\alpha|)\delta} \, D^{\alpha}_{\underline{y}} \, v(\underline{y}) \in L^2(\mathbb{R}^n_{\underline{y}}) \quad \text{for every } \alpha \text{ such that } |\alpha| \leq m .$$

Grushin proves the following results:

Proposition 2.21. If $L(\underline{y},D_{\underline{x}},D_{\underline{y}})$ of type (2.45) has quasi-homogeneous symbol and satisfies Condition 1 then for every $\underline{\xi} \neq 0$ the partial Fourier transformed operator

$$(2.49) \quad L(\underline{y}, \underline{\xi}, D_{\underline{y}}) : H_{(m,\delta)}(\mathbb{R}^n_{\underline{y}}) \longrightarrow L^2(\mathbb{R}^n_{\underline{y}})$$

is Fredholm.

Proposition 2.22. Under the same hypotheses as above, for any $\underline{\xi} \neq 0$ any solution $v(\underline{y})$ of $L(\underline{y}, \underline{\xi}, D_{\underline{y}})\, v(\underline{y}) = 0$ in $L^2(\mathbb{R}^n_{\underline{y}})$ lies in $\mathcal{S}(\mathbb{R}^n_{\underline{y}})$, the Schwartz space.

As a corollary of Proposition 2.22 he gets

Corollary 2.23. Under the same hypotheses as above, the following conditions are all equivalent:

(2.50) $L(\underline{y}, \xi, D_{\underline{y}})\ v(\underline{y}) = 0$ has a non-trivial solution in $\mathcal{S}$.

(2.51) $\dim \operatorname{Ker} L(\underline{y}, \xi, D_{\underline{y}}) \neq 0$, where $L(\underline{y}, \xi, D)$ is viewed as an operator from $H_{(m,\delta)}(\mathbb{R}^n_{\underline{y}})$ to $L^2(\mathbb{R}^n_{\underline{y}})$.

(2.52) Zero is an eigenvalue of $L(\underline{y}, \xi, D_{\underline{y}})$, where $L(\underline{y}, \xi, D_{\underline{y}})$ is viewed as an unbounded operator on $L^2(\mathbb{R}^n_{\underline{y}})$ with domain $H_{(m,\delta)}(\mathbb{R}^n_{\underline{y}})$.

He shows also

Proposition 2.24. Under the same hypotheses as above, let $v(\underline{y})$ be a tempered distribution satisfying the equation

(2.53) $L(\underline{y}, \xi, D_{\underline{y}})\ v(\underline{y}) = f(\underline{y})$.

If $f \in \mathcal{S}$ then $v \in \mathcal{S}$, and if $f \in L^2(\mathbb{R}^n_{\underline{y}})$, then $v \in H_{(m,\delta)}(\mathbb{R}^n_{\underline{y}})$.

By applying Prop. 2.24 to $L^*(\underline{y}, D_{\underline{x}}, D_{\underline{y}})$, the formal adjoint of $L(\underline{y}, D_{\underline{x}}, D_{\underline{y}})$ (which satisfies the same hypotheses as $L(\underline{y}, D_{\underline{x}}, D_{\underline{y}})$) we see that the Hilbert space adjoint, $[L(\underline{y}, \underline{\xi}, D_{\underline{y}})]^*$, of $L(\underline{y}, \underline{\xi}, D_{\underline{y}})$ has domain $H_{(m,\delta)}(R^n y)$. Hence, it follows that $[L(\underline{y}, \underline{\xi}, D_{\underline{y}})]^* = L^*(\underline{y}, \underline{\xi}, D_{\underline{y}})$, the partial Fourier transform of $L^*(\underline{y}, D_{\underline{x}}, D_{\underline{y}})$. In particular we have

<u>Corollary 2.25.</u> Ker $L^*(\underline{y}, \underline{\xi}, D_{\underline{y}}) = [\text{Range } L(\underline{y}, \underline{\xi}, D_{\underline{y}})]^{\perp}$. Thus, dim Ker $L^*(\underline{y}, \underline{\xi}, D_{\underline{y}})$ = dim Coker $L(\underline{y}, \underline{\xi}, D_{\underline{y}})$.

In §2.4 we shall refer to results of Grushin on the index of the Fredholm operators $L(\underline{y}, \underline{\xi}, D_{\underline{y}})$. For our present purposes we need the following result of Grushin.

<u>Theorem 2.26</u> Suppose that $L(\underline{y}, D_{\underline{x}}, D_{\underline{y}})$ is of type (2.45), has quasi-homogeneous symbol, and satisfies Condition 1. Then $L(\underline{y}, D_{\underline{x}}, D_{\underline{y}})$ is hypoelliptic if and only if for every $\underline{\xi}$ such that $|\underline{\xi}| = 1$ $L(\underline{y}, \underline{\xi}. D_{\underline{y}})$ does <u>not</u> have 0 as an eigenvalue, that is, if and only if Ker $L(\underline{y}, \underline{\xi}, D_{\underline{y}}) = \{0\}$.

Although we shall not need this result, we mention that Grushin also proves

<u>Theorem 2.27.</u> Under the same hypotheses as above, if for every $\underline{\xi}$ such that $|\underline{\xi}| = 1$ Coker $L(\underline{y}, \underline{\xi}, D_{\underline{y}}) = \{0\}$, then $L(\underline{y}, D_{\underline{x}}, D_{\underline{y}})$ is locally solvable.

I do not know whether the condition of Theorem 2.27 is necessary as well as sufficient for local solvability.

Returning to the auxiliary operator $\check{P}_{(x_o, \xi_o)}(\underline{s}, D_{\underline{s}}, D_r)$ defined by (2.44), we shall see that if we take $m = 2$ and $\delta = 1$, and take $\underline{y} = \underline{s}$ and $\underline{x} = r$, then $\check{P}_{(x_o, \xi_o)}(\underline{s}, D_{\underline{s}}, D_r)$ is of type (2.45) and is quasi-homogeneous and satisfies Condition 1.

Indeed, let $\check{p}_{(x_o, \xi_o)}(\underline{s}, \underline{\eta}, \rho)$ denote the total symbol of $\check{P}_{(x_o, \xi_o)}(\underline{s}, D_{\underline{s}}, D_r)$ with respect to the coordinates $\underline{s}$, r and corresponding cotangential coordinates $\underline{\eta}$, ρ. Then

$$(2.54) \quad \check{p}_{(x_o, \xi_o)}(\underline{s}, \underline{\eta}, \rho) = \Sigma\, a_{ij}|_{(x_o, \xi_o)}\, \eta_i\, \eta_j + \Sigma\, b_{ij}|_{(x_o, \xi_o)}\, s_i\, s_j\, \rho^2 + \Sigma\, 2\, c_{ij}\, s_i\, \eta_j \rho + \left(\sigma_{sub}(P)|_{(x_o, \xi_o)} + \frac{1}{\sqrt{-1}} \sum_i c_{ii}|_{(x_o, \xi_o)}\right) \rho \ .$$

Hence one sees easily that for every $\lambda > 0$

$$(2.55) \quad \check{p}_{(x_o, \xi_o)}(\underline{s}/\lambda, \lambda\underline{\eta}, \lambda^2 \rho) = \lambda^2\, \check{p}_{(x_o, \xi_o)}(\underline{s}, \underline{\eta}, \rho) \ .$$

Thus, the quasi-homogeneity condition is satisfied.

To see that Condition 1 is satisfied recall from (2.7) that $\tilde{p}_{(x_o, \xi_o)}(v, v) = 0 \Longrightarrow v = 0$, where $v \in N(\Sigma)_{(x_o, \xi_o)}$. Thus,

$$(2.56)\quad \overset{\vee o}{P}_{(x_o,\xi_o)}(\underline{s}, \underline{\eta}, \rho) \equiv \Sigma\, a_{ij}|_{(x_o,\xi_o)}\, \eta_i\, \eta_j + \Sigma\, b_{ij}|_{(x_o,\xi_o)}\, s_i\, s_j\, \rho^2 + \Sigma\, 2\, c_{ij}|_{(x_o,\xi_o)}\, s_i\, \eta_j\, \rho \neq 0$$

for $\underline{s} \neq 0$ and for $|\underline{\eta}|+|\rho| \neq 0$.

In fact to prove this just use (2.7) and (2.42)(or(2.41)) and the fact that $\underline{v} \equiv (\rho\underline{s}, \underline{\eta}) \neq 0$ if $\underline{s} \neq 0$ and $|\eta|+|\rho| \neq 0$. This shows that Condition 1 holds.

Since all of Grushin's hypotheses hold we can apply Theorem 2.26 and get

(2.57) $\overset{\vee}{P}_{(x_o,\xi_o)}(\underline{s}, D_{\underline{s}}, D_r)$ is hypoelliptic $\Longleftrightarrow$

$\overset{\vee}{P}_{(x_o,\xi_o)}(\underline{s}, D_{\underline{s}}, 1)$ and $\overset{\vee}{P}_{(x_o,\xi_o)}(\underline{s}, D_{\underline{s}}, -1)$ both fail to have 0 as an eigenvalue .

(Here we have taken the partial Fourier transform with respect to r and then evaluated at the two points $\rho = \pm 1$.)

But observe that

$$(2.58)\quad \overset{\vee}{P}_{(x_o,\xi_o)}(\underline{s}, D_{\underline{s}}, 1) = \Sigma\, a_{ij}|_{(x_o,\xi_o)}\, D_{s_i} D_{s_j} + \Sigma\, b_{ij}|_{(x_o,\xi_o)}\, s_i\, s_j + \Sigma\, c_{ij}|_{(x_o,\xi_o)}\, s_i\, D_{s_j} + \Sigma\, c_{ij}|_{(x_o,\xi_o)}\, D_{s_i} s_i + \sigma_{sub}(P)|_{(x_o,\xi_o)}$$

$$= \widetilde{P}_{(x_o,\xi_o)} + \sigma_{sub}(P)|_{(x_o,\xi_o)}$$

and

$$(2.59)\quad \check{P}_{(x_o,\xi_o)}(\underline{s}, D_{\underline{s}}, -1) = \Sigma\, a_{ij}|_{(x_o,\xi_o)}\, D_{s_i} D_{s_j} +$$

$$\Sigma\, b_{ij}|_{(x_o,\xi_o)}\, s_i\, s_j - \Sigma\, c_{ij}|_{(x_o,\xi_o)}\, s_i D_{s_j} -$$

$$\Sigma\, c_{ij}|_{(x_o,\xi_o)}\, D_{s_j}\, s_i - \sigma_{sub}(P)|_{(x_o,\xi_o)}$$

Remark 2.28. Suppose our original operator P had its principal symbol and sub-principal part fully homogeneous with respect to $\mathbb{R}$ rather than just with respect to $\mathbb{R}^+$, as would be the case if, for example, P were a differential operator. In particular, Σ would be conic with respect to $\mathbb{R}$. Hence, if we chose as local parameters for Σ at (x_o,ξ_o) (satisfying (2.37)) functions $u_1,\dots,u_n$ homogeneous of degree 0 (with respect to $\mathbb{R}$) and functions $v_1,\dots,v_n$ homogeneous of degree 1, then $u_1,\dots,u_n$, $v_1,\dots,v_n$ also could be used as local parameters (satisfying (2.37)) for Σ at $(x_o,-\xi_o)$. Using as coordinates in $N(\Sigma)_{(x_o,\xi_o)}$ and $N(\Sigma)_{(x_o,-\xi_o)}$ the $s_1,\dots,s_n$, $t_1,\dots,t_n$ given by (2.38) we would see from (2.59) and from the $\mathbb{R}$-homogeneity of p and $\sigma_{sub}(P)|_\Sigma$ that

$$(2.60)\quad \check{P}_{(x_o,\xi_o)}(\underline{s}, D_{\underline{s}}, -1) = (-1)^m\, [\tilde{P}_{(x_o,-\xi_o)} + \sigma_{sub}(P)_{(x_o,-\xi_o)}]$$

Hence, (2.57) could be restated as : $\check{P}_{(x_o,\xi_o)}(\underline{s}, D_{\underline{s}}, D_r)$ is hypoelliptic if and only if neither $\widetilde{P}_{(x_o,\xi_o)} + \sigma_{sub}(P)_{(x_o,\xi_o)}$ nor $\widetilde{P}_{(x_o,-\xi_o)} + \sigma_{sub}(P)_{(x_o,-\xi_o)}$ has 0 as an eigenvalue. Using this fact we could avoid the use of the microlocalized version (2.68), below, of (2.57) which we need in proving Theorem 2.4 in the general case of $\mathbb{R}^+$ - homogeneity.

We next define a microlocalized version of hypoellipticity with loss of 1 derivative.

<u>Definition 2.29.</u> $Q(\underline{x}, D_{\underline{x}})$ of order r is <u>hypoelliptic at $(x_1, \xi_1) \in T^*X/\{0\}$ with loss of one derivative</u> $\Longleftrightarrow$ there exists a conic open neighborhood Γ of (x_1, ξ_1) such that

(2.61) For every distribution u , if $Qu \in H^s_{loc}$ <u>and</u> $WF(u) \subset \Gamma$, then $u \in H^{s+r-1}_{loc}$. (Here $WF(u)$ denotes the wave-front set of u .)

It suffices to consider (2.61) only at points (x_1, ξ_1) in the characteristic variety of Q , for standard techniques allow us to show that if Q is "elliptic at (x_1, ξ_1)" then Q is hypoelliptic at (x_1, ξ_1) with no loss of derivatives, i.e. (2.61) holds with r replacing $r - 1$. From Boutet de Monvel and Treves ([3] , [4]) we know that an operator Q satisfying the conditions of

Theorem 2.4 (including the "conflicting influences" condition (see Remark 2.2A) in case $n = 1$) is hypoelliptic with loss of 1 derivative at $(x, \xi) \in$ characteristic variety of Q if and only if a certain criterion $B - T_{(x,\xi)}$ <u>fails to hold</u> at each point (x, ξ) in some conic neighborhood of (x_1, ξ_1) in the characteristic variety. We shall not need to make use here of the explicit terms of the criterion $B - T_{(x,\xi)}$, but rather its general features. Although the condition $B - T_{(x,\xi)}$ is not stated in [4] in terms of $\tilde{q}_{(x,\xi)}$ and $\sigma_{sub}(Q)|_{(x,\xi)}$ (where q denotes the principal symbol of Q), it is easy to check, in view of (2.34), that $B - T_{(x,\xi)}$ depends only on knowing the expression of $\tilde{q}_{(x,\xi)}$ in terms of a set of symplectic linear coordinates for $N(\Sigma)_{(x,\xi)}$ and on knowing $\sigma_{sub}(Q)_{(x,\xi)}$. (We remark that although the Z_o occurring in [4] is not the same as our S occurring in (2.34), $\sigma_{r-1}(Z_o)_{(x,\xi)}$ differs from $\sigma_{r-1}(S)_{(x,\xi)}$ by terms involving Poisson brackets <u>at (x, ξ)</u> of local parameters defining Σ, and these terms are uniquely determined by the expression of $\tilde{q}_{(x,\xi)}$ in terms of symplectic linear coordinates.

From the explicit description of condition $B - T_{(x,\xi)}$ in [4] it is easy to check, via continuity considerations, that if $B - T_{(x_1,\xi_1)}$ fails to hold, then $B - T_{(x,\xi)}$ fails to hold for (x, ξ) in a conic neighborhood of (x_1, ξ_1)

in the characteristic variety. Thus, we know that, under the assumption that Q satisfies the hypotheses of Theorem 2.4,

(2.62) Q is hypoelliptic with loss of 1 derivative at $(x_1, \xi_1) \in$ characteristic variety $\Longleftrightarrow B - T_{(x_1, \xi_1)}$ fails to hold .

Now, in view of the relationship of P and $\check{P}_{(x_0, \xi_0)}$ (compare (2.40) with (2.42) and (2.43)), and because the hypotheses of Theorem 2.4 (including the "conflicting influences" asumption) involve only the principal symbol of the operator in question, it follows that since P satisfies these hypotheses, so does $\check{P}_{(x_0, \xi_0)}$ for any $(x_0, \xi_0) \in \Sigma$. Hence, observing (see (2.56)) that the characteristic variety of $\check{P}_{(x_0, \xi_0)}$ (for fixed (x_0, ξ_0)) is $\{(\underline{s}, r, \underline{\eta}, \rho) \mid \underline{s} = 0$ and $\underline{\eta} = 0\}$, we apply (2.62) to $\check{P}_{(x_0, \xi_0)}$ and get

(2.63) $\check{P}_{(x_0, \xi_0)}(\underline{s}, D_{\underline{s}}, D_r)$ is hypoelliptic with loss of 1 derivative at $(\underline{0}, r_1, \underline{0}, \rho_1) \Longleftrightarrow B - T_{(\underline{0}, r_1, \underline{0}, \rho_1)}$ fails to hold.

Since none of the coefficients of $\check{P}_{(x_0, \xi_0)}(\underline{s}, D_{\underline{s}}, D_r)$ involve r it follows that $B - T_{(\underline{0}, r, \underline{0}, \rho)}$ does not depend on r. Hence, since $B - T_{(x_0, \rho\xi_0)}$ holds for all

$\rho > 0$ if and only if $B - T_{(x_0, \xi_0)}$ holds, it follows that

(2.64) $\check{P}_{(x_0,\xi_0)}(\underline{s}, D_{\underline{s}}, D_r)$ is hypoelliptic with loss of 1 derivative at every point $(\underline{0}, r, \underline{0}, \rho)$ with $\rho > 0 \Longleftrightarrow B - T_{(\underline{0}, 0, \underline{0}, 1)}$ fails to hold.

But Boutet de Monvel and Treves ([3], [4]) show that these pointwise results can be "patched together". More precisely, they show that for Q satisfying the hypotheses of Theorem 2.4

(2.65) Q is hypoelliptic with loss of 1 derivative at each point of the conic neighborhood Γ in $T^*X/\{0\} \Longleftrightarrow$ for every distribution u, if $Qu \in H^s_{loc}$ and $WF(u) \subset \Gamma$, then $u \in H^{s+r-1}_{loc}$.

(Notice that (2.65) is not obvious. The condition in (2.65) seems stronger a priori than hypoellipticity with loss of 1 derivative at each point of Γ, for $WF(u)$ is not required to lie in a smaller conic neighborhood than Γ varying with the point in question.)

Combining (2.64) and (2.65) we conclude that

(2.66) $\check{P}_{(x_0,\xi_0)}(\underline{s}, D_{\underline{s}}, D_r)$ satisfies the following "semi-microlocalized" hypoellipticity property (2.67) $\Longleftrightarrow B - T_{(\underline{0}, 0, \underline{0}, 1)}$ fails to hold.

(2.67) For every distribution u whose WF set is contained in $\{(\underline{s}, r, \underline{\eta}, \rho) | \rho > 0\}$ $\check{P}_{(x_o, \xi_o)}(\underline{s}, D_{\underline{s}}, D_r)u \in H^t_{loc} \Longrightarrow u \in H^{t+1}_{loc}$.

(Remember that $\check{P}_{(x_o, \xi_o)}$ is of order 2 .)

We shall also want to use the following semi-microlocalized version of (2.57) .

(2.68) $\check{P}_{(x_o, \xi_o)}(\underline{s}, D_{\underline{s}}, D_r)$ satisfies the semi-microlocalized hypoellipticity property (2.67) $\Longleftrightarrow P_{(x_o, \xi_o)}(\underline{s}, D_{\underline{s}}, 1)$ fails to have 0 as an eigenvalue.

(The idea is that the restriction to those distributions u whose WF sets do not meet any points in the cotangent space with $\rho < 0$ allows us to get rid precisely of the condition in (2.57) involving the partial Fourier transform of $\check{P}_{(x_o, \xi_o)}$ with respect to r at points where $\rho < 0$.) To prove (2.68) we examine Grushin's proof ([13]) of Theorem 2.26 , from which (2.57) followed. Grushin proves that the eigenvalue condition is necessary for hypoellipticity by showing how to construct, in case 0 is an eigenvalue of $L(\underline{y}, \underline{\xi}^o, D_{\underline{y}})$ for some $\underline{\xi}^o$, a distribution u such that $L(\underline{y}, D_{\underline{x}}, D_{\underline{y}})u = 0$ but such that u is not in C^∞ . It is not hard to check that the distribution u he constructs satisfies:

$WF(u) \subset \{(\underline{x}, \underline{0}, \rho\underline{\xi}^{\circ}, \underline{\eta}) \mid \rho > 0\}$. This proves, in particular, the implication ($\Longleftarrow$) of (2.68). As mentioned at the beginning of this section, we have not tried to carry out the details of the argument sketched below for the opposite implication ($\Longrightarrow$) of (2.68), but we see no reason why this cannot be done. Grushin proves the sufficiency of the eigenvalue condition for hypoellipticity by constructing a left parametrix R for $L(\underline{y}, D_{\underline{x}}, D_{\underline{y}})$ with good regularity properties. Roughly speaking, he constructs R by taking for each $\underline{\xi} \neq 0$ a left-inverse $R(\underline{y}, \underline{\xi}, D_{\underline{y}})$ of $L(\underline{y}, \underline{\xi}, D_{\underline{y}})$ and then defining $(Ru)(\underline{x}, \underline{y})$ as $\int e^{i\underline{x} \cdot \underline{\xi}} R(\underline{y}, \underline{\xi}, D\underline{y})\, \tilde{u}(\underline{\xi}, \underline{y})\, d\underline{\xi}$, where $\tilde{u}$ denotes partial Fourier transform with respect to $\underline{x}$. Introducing certain weighted Sobolev spaces, he proves a regularity property for R which implies, we believe, that $L(\underline{y}, D_{\underline{x}}, D_{\underline{y}})$ is hypoelliptic with _gain_ of $\frac{m}{1+\delta}$ derivatives. (_Gain_ is defined as (m-_loss_).) In our case, with $\check{P}_{(x_0, \xi_0)}(\underline{s}, D_{\underline{s}}, D_r)$ as $L(\underline{y}, D_{\underline{x}}, D_{\underline{y}})$, $\underline{\xi}$ is replaced by ρ. Assuming that $\check{P}_{(x_0, \xi_0)}(\underline{s}, D_{\underline{s}}, 1)$ does not have 0 as an eigenvalue, we replace the above R by $(Ru)(\underline{s}, r) = \int_0^\infty e^{ir\rho} R(\underline{s}, \rho, D_{\underline{s}})\, \tilde{u}(\rho, \underline{s})\, d\rho$. (Note that we integrate only over the positive half-line, rather than over the whole line.) Grushin's argument can now, most likely, be modified to yield the implication ($\Longleftarrow$) of (2.68). We assume now that (2.68) has in fact been proved.

As we saw in (2.58), $\check{P}_{(x_0, \xi_0)}(\underline{s}, D_{\underline{s}}, 1) = \tilde{P}_{(x_0, \xi_0)} + \sigma_{sub}(P)_{(x_0, \xi_0)}$. Hence, combining (2.66) and (2.68), we see that

$$(2.69) \qquad B - T_{(\underline{0},\ 0,\ \underline{0},\ 1)} \text{ fails to hold for } \check{P}_{(x_0, \xi_0)}(\underline{s}, D_{\underline{s}}, D_r) \Longleftrightarrow \tilde{P}_{(x_0,\xi_0)} + \sigma_{sub}(P)_{(x_0,\xi_0)} \text{ fails to have } 0 \text{ as an eigenvalue.}$$

But, as we stated earlier, conditon $B - T_{(x,\xi)}$ for Q depends only on knowing the expression of $\tilde{q}_{(x,\xi)}$ in terms of a set of symplectic linear coordinates for $N(\Sigma)_{(x,\xi)}$ and on knowing $\sigma_{sub}(Q)|_{(x,\xi)}$. Hence, comparing (2.40) with (2.42) and (2.43) , and checking that (for any r)

$$(2.70) \qquad \sigma_{sub}(\check{P}_{(x_0,\xi_0)})|_{(\underline{0},\ r,\ \underline{0},\ 1)} = \sigma_{sub}(P)|_{(x_0,\xi_0)}$$

we see that

$$(2.71) \qquad B - T_{(\underline{0},\ 0,\ \underline{0},\ 1)} \text{ holds for } \check{P}_{(x_0,\xi_0)}(\underline{s}, D_{\underline{s}}, D_r) \Longleftrightarrow B - T_{(x_0,\xi_0)} \text{ holds for } P .$$

In view of (2.69) this becomes

$$(2.72) \qquad B - T_{(x,\xi)} \text{ fails to hold for } P \text{ at } (x,\xi) \in \Sigma \Longleftrightarrow \tilde{P}_{(x,\xi)} + \sigma_{sub}(P)|_{(x,\xi)} \text{ fails to have } 0 \text{ as an eigenvalue.}$$

In particular, in view of (2.62) and (2.65) , Theorem 2.4 follows. (We have tacitly assumed, for ease of presentation, that the lower order part of P , as well as the leading part, is scalar. The same discussion work equally well in the general case, where the lower order part of P has $N\times N$ matrix coefficients.)

§2.4. The index of the test-operators

In order to use the results of [4] in the proof of Theorem 2.4 we had to assume, in the case $n = 1$, the "conflicting influences" condition:

(2.73) The winding number of p about $\Sigma = 0$.

(See Remark 2.2A). This condition corresponds, in the "abstract set-up" (1.3) of §1. to the case where $\operatorname{Re} a_o'(0)$ and $\operatorname{Re} b_o'(0)$ have opposite sign. The name "conflicting influences" comes from the fact [25] that if $\operatorname{Re} a_o'(0)$ and $\operatorname{Re} b_o'(0)$ have opposite sign, then one of the first-order factors in (1.3) is hypoelliptic but not locally solvable and the other factor is locally solvable but not hypoelliptic. It is proved in [25] that under the "conflicting influences" assumption, P in (1.3) is hypoelliptic if and only if it is locally solvable, and that, furthermore, these two equivalent conditions will hold provided a set of discrete conditions involving $c(t,A)A$ fail to hold. In case $\operatorname{Re} a_o'(0)$ and $\operatorname{Re} b_o'(0)$ are both <0 then P is automatically locally solvable and not hypoelliptic, and in case $\operatorname{Re} a_o'(0)$ and $\operatorname{Re} b_o'(0)$ are both >0, P is hypoelliptic but is not locally solvable. Thus, in case $\operatorname{Re} a_o'(0)$ and $\operatorname{Re} b_o'(0)$ have the same sign the local solvability and hypoellipticity of P are mutually exclusive, and, furthermore, the lower order part of P

does not play a role as far as hypoellipticity or local solvability is concerned. Also, no discrete set of conditions arise.

Returning to the context of Theorem 2.4 we assume that a Riemannian metric for the base manifold and a Hermitian metric along the fibers of the vector bundle have been chosen, so that P^*, the formal adjoint of P, is defined. It is shown in [4] that under the assumptions of Theorem 2.4, including the "conflicting influences" condition in case $n = 1$, P is hypoelliptic with loss of 1 derivative if and only if P^* is hypoelliptic with loss of 1 derivative, this latter condition implying that P is locally solvable. Also the lower order part of P is involved and a set of discrete conditions arises.

We shall see that this behavior has a natural interpretation in terms of the index of the test-operators $\widetilde{P}_{(x,\xi)} + \sigma_{sub}(P)|_{(x,\xi)}$. The deviation of local solvability of P from hypoellipticity of P is measured by this index (more precisely, the deviation of hypoellipticity of P^* from hypoellipticity of P). Moreover, this index, which does not depend on the lower order part of P, determines whether $\widetilde{P}_{(x,\xi)} + \sigma_{sub}(P)|_{(x,\xi)}$ has all of $\mathbb{C}$, the null set, or a countable set without limit points as its set of eigenvalues. This accounts for the absence or presence of discrete conditions in the criteria for hypoellipticity and local solvability of P.

The discussion below indicates strongly that Theorem 2.4 holds even in the case $n = 1$ without the further assumption of "conflicting influences". However, the condition that $\tilde{P}_{(x,\xi)} + \sigma_{sub}(P)|_{(x,\xi)}$ not have 0 as an eigenvalue will be purely a hypoellipticity criterion and not a local solvability criterion. In fact, it will be a non-hypoellipticity criterion for P^*. Also, the lower order part of P will not play a role, and no discrete conditions will arise.

If we define P <u>locally solvable with loss of 1 derivative</u> to mean that P^* is hypoelliptic with loss of 1 derivative, then we have the following local solvability theorem corresponding to the hypoellipticity theorem 2.4 .

<u>Theorem 2.4'</u>. P satisfying the hypotheses of Theorem 2.4 is locally solvable with loss of one derivative if and only if for every $(x,\xi) \in \Sigma$, $\tilde{P}_{(x,\xi)} + \sigma_{sub}(P)|_{(x,\xi)} : H_{(2,1)} \longrightarrow L^2$ has $\{0\}$ cokernel, i.e. is surjective.

We shall see that this theorem is certainly true for $n > 1$ and is true for $n = 1$ under the "conflicting influences" assumption. Furthermore, it will be true in the non-conflicting influences case provided that Theorem 2.4 holds in this case, as we believe it does. In this case it will be a non-hypoellipticity criterion for P. Furthermore, the lower order part of P will not play a role and no discrete conditions will arise.

We begin by discussing some results of Grushin ([14]) on the index of the Fredholm operators $L(\underline{y}, \underline{\xi}, D_{\underline{y}})$: $H_{(m,\delta)}(\mathbb{R}^n_{\underline{y}}) \longrightarrow L^2(\mathbb{R}^n_{\underline{y}})$, i.e.,

$$(2.74) \qquad \text{ind } L_{\underline{\xi}} \equiv \dim\ker L(\underline{y}, \underline{\xi}, D_{\underline{y}}) - \dim\operatorname{coker} L(\underline{y}, \underline{\xi}, D_{\underline{y}}) .$$

(We shall use the same notation as in our earlier discussion of Grushin's results beginning with (2.45) . As before, $n = \frac{1}{2}$ codimension of characteristic variety. In the case of Grushin's operators this corresponds to the number of $\underline{y}$ variables. Also, we let k be the number of $\underline{x}$ variables.) Grushin shows

Proposition 2.30. Let $L(\underline{y}, D_{\underline{x}}, D_{\underline{y}})$ have scalar principal symbol and satisfy the hypotheses of Proposition 2.21. Then

(2.75) ind $L_{\underline{\xi}}$ is independent of $\underline{\xi} \neq 0$ if $k > 1$, and depends only on the sign of ξ if $k = 1$. (This follows directly from the homotopy invariance of the index.)

(2.76) If $n > 1$ then for every $\underline{\xi} \neq 0$, ind $L_{\underline{\xi}} = 0$.

(2.77) If $n = 1$ then ind $L_{\underline{\xi}} = \nu_+ + \nu_- - m$, where

ν_+ = the number of roots of the equation $L^o(1, \underline{\xi}, \zeta) = 0$ for which $\operatorname{Im} \zeta > 0$, and ν_- = the number of roots of the equation $L^o(-1, \underline{\xi}, \zeta) = 0$ for which $\operatorname{Im} \zeta < 0$. Here $L^o(\underline{y}, \underline{\xi}, \underline{n})$ denotes the principal symbol of $L(\underline{y}, D_{\underline{x}}, D_{\underline{y}})$, and m denotes the order of $L(\underline{y}, D_{\underline{x}}, D_{\underline{y}})$.

Notice that ind $L_{\underline{\xi}}$ depends only on the principal symbol L^o of $L(\underline{y}, D_{\underline{x}}, D_{\underline{y}})$, this being, of course , another consequence of the homotopy invariance of the index.

Now, take $\check{P}_{(x_o, \xi_o)}(\underline{s}, D_{\underline{s}}, D_r)$ as the Grushin operator $L(\underline{y}, D_{\underline{x}}, D_{\underline{y}})$ (with $m = 2$, $\delta = 1$), and recall (2.58) that $\check{P}_{(x_o, \xi_o)}(\underline{s}, D_{\underline{s}}, 1) = \tilde{P}_{(x_o, \xi_o)} + \sigma_{sub}(P)|_{(x_o, \xi_o)}$. It then follows from the preceding proposition that

<u>Proposition 2.31.</u> If $n > 1$ the index of $\tilde{P}_{(x_o, \xi_o)} + \sigma_{sub}(P)|_{(x_o, \xi_o)} = 0$. If $n = 1$ the index of $\tilde{P}_{(x_o, \xi_o)} + \sigma_{sub}(P)|_{(x_o, \xi_o)} = \nu_+ + \nu_- - 2$ where (see(2.56)) ν_+ = the number of roots ζ with $\operatorname{Im} \zeta > 0$ of the equation

$$a|_{(x_o, \xi_o)}\zeta^2 + 2c|_{(x_o, \xi_o)}\zeta + b|_{(x_o, \xi_o)} = 0 \tag{2.78}$$

and ν_- = the number of roots ζ with $\operatorname{Im} \zeta < 0$ of the equation

$$a|_{(x_o, \xi_o)}\zeta^2 - 2c|_{(x_o, \xi_o)}\zeta + b|_{(x_o, \xi_o)} = 0 \ . \tag{2.79}$$

(Remember that n for $\check{P}_{(x_o,\xi_o)}(\underline{s}, D_{\underline{s}}, D_r)$ is the same as the n associated to P, i.e., $\frac{1}{2}$ codim Σ. Since, for $n = 1$, i,j both must equal 1 in (2.56) we have simply suppressed these indices in (2.78) and (2.79). Notice also that the index depends only on $\tilde{P}_{(x_o,\xi_o)}$, i.e., only on $\tilde{p}_{(x_o,\xi_o)}$, and not at all on $\sigma_{sub}(P)|_{(x_o,\xi_o)}$.)

By using (2.34) it is easy to verify that

$$\text{(2.80a)} \qquad \sigma_{sub}(P^*)|_\Sigma = (\sigma_{sub}(P)|_\Sigma)^*$$

and, hence, that

$$\text{(2.80b)} \qquad (P^*)^{\vee}_{(x_o,\xi_o)} = (\check{P}_{(x_o,\xi_o)})^*$$, the formal adjoint of $\check{P}_{(x_o,\xi_o)}$.

This gives, in view of Cor. 2.25,

$$\text{(2.81)} \qquad \dim\ker(\widetilde{P^*}_{(x_o,\xi_o)} + \sigma_{sub}(P^*)|_{(x_o,\xi_o)}) = \dim\operatorname{coker}(\tilde{P}_{(x_o,\xi_o)} + \sigma_{sub}(P)|_{(x_o,\xi_o)}) .$$

Thus, by Prop 2.31,

(2.82) If $n > 1$, 0 is an eigenvalue of $\tilde{P}_{(x_o,\xi_o)} + \sigma_{sub}(P)|_{(x_o,\xi_o)}$ if an only if 0 is an eigenvalue of $\widetilde{P^*}_{(x_o,\xi_o)} + \sigma_{sub}(P^*)|_{(x_o,\xi_o)}$.

But in the case $n > 1$ we know that Theorem 2.4 holds (both for P and for P^*). Thus,

(2.83) If $n > 1$, then P is hypoelliptic with loss of 1 derivative if and only if P^* is hypoelliptic with loss of 1 derivative.

Let us next analyze more closely the case $n = 1$. The roots of equation (2.78) are $\dfrac{-c \pm \sqrt{c^2 - ab}}{a}$ and those of equation (2.79) are $\dfrac{c \pm \sqrt{c^2 - ab}}{a}$, i.e., if the roots of equation (2.78) are denoted ζ_1, ζ_2, then those of equation (2.79) are $-\zeta_1, -\zeta_2$.

Thus, $\nu_+ = \nu_-$. But (2.7) implies that equation (2.78) has no real root, for if there were a real root ζ, then the real vector $(\zeta, 1)$ would satisfy:

$$a|_{(x_0,\xi_0)}\zeta^2 + 2c|_{(x_0,\xi_0)}\,\zeta \cdot 1 + b|_{(x_0,\xi_0)}\,1 \cdot 1 = 0$$

Hence, using Prop 2.31, we see that the following proposition gives a complete enumeration of the possible behavior of $\operatorname{ind}\left(\tilde{P}_{(x_0,\xi_0)} + \sigma_{sub}(P)|_{(x_0,\xi_0)}\right)$ in the case $n = 1$.

Proposition 2.32.

(2.84) If both roots of (2.78) have positive imaginary part, $\operatorname{ind}(\tilde{P}_{(x_o,\xi_o)} + \sigma_{sub}(P)|_{(x_o,\xi_o)}) = 2$.

(2.85) If both roots of (2.78) have negative imaginary part, $\operatorname{ind}(\tilde{P}_{(x_o,\xi_o)} + \sigma_{sub}(P)|_{(x_o,\xi_o)}) = -2$.

(2.86) If the two roots of (2.78) have opposite sign, $\operatorname{ind}(\tilde{P}_{(x_o,\xi_o)} + \sigma_{sub}(P)|_{(x_o,\xi_o)}) = 0$.

But (see [25] and [3]) the winding number of p about Σ at (x_o,ξ_o) is given as follows: write p locally about (x_o,ξ_o) as $p = av^2 + 2cuv + bu^2$, where u, v are local parameters defining Σ . Then, under the hypothesis that $a\zeta^2 + 2c\zeta + b = 0$ has no real roots ζ , the winding number of p about Σ at (x_o,ξ_o) is defined as the index i_p of the mapping from $\mathbb{C} - \{0\}$ into $\mathbb{C} - \{0\}$ given by

$$(2.87) \qquad x + iy \longmapsto a|_{(x_o,\xi_o)}\, y^2 + 2c|_{(x_o,\xi_o)}\, xy + b|_{(x_o,\xi_o)}\, x^2 \, .$$

Moreover, $i_p = 2$ in case both roots ζ_1, ζ_2 of

$a|_{(x_o,\xi_o)}\zeta^2 + 2c|_{(x_o,\xi_o)}\ \zeta + b|_{(x_o,\xi_o)} = 0$ have positive imaginary part; $i_p = -2$ if both roots have negative imaginary part ; and $i_p = 0$ in case the two roots have opposite sign. Since the equation in question is just (2.78) it follows from Prop 2.32 that

(2.38) If $n = 1$, then the winding number of p about Σ at (x_o,ξ_o) equals $\text{ind}\,(\tilde{P}_{(x_o,\xi_o)} + \sigma_{sub}(P)|_{(x_o,\xi_o)})$.

In particular, the case of "conflicting influences" corresponds precisely to the condition that $\text{ind}(\tilde{P}_{(x_o,\xi_o)} + \sigma_{sub}(P)|_{(x_o,\xi_o)}) = 0$. Since we know Theorem 2.4 holds in this case, we can argue exactly as in (2.83) and show

(2.89) If $n = 1$ and $\text{ind}(\tilde{P}_{(x_o,\xi_o)} + \sigma_{sub}(P)|_{(x,\xi)}) = 0$ for every $(x,\xi) \in \Sigma$, then P is hypoelliptic with loss of 1 derivative if and only if P^* is hypoelliptic with loss of 1 derivative.

Consider now the cases where $\text{ind}(\tilde{P}_{(x,\xi)} + \sigma_{sub}(P)|_{(x,\xi)})$ $\neq 0$. If $\text{ind} > 0$ then $\dim\ker > 0$. Since the index depends only on $\tilde{P}_{(x,\xi)}$ and not on the lower order part, we see from (2.68) that if $\text{ind} > 0$, then $\check{P}_{(x,\xi)}(\underline{s}, D_{\underline{s}}, D_r)$

automatically fails to satisfy the semi-microlocalized hypoellipticity condition (2.67), regardless of what $\sigma_{sub}(P)|_{(x,\xi)}$ is (and as we see by also using (2.91) below, $\check{P}^*_{(x,\xi)}(\underline{s}, D_{\underline{s}}, D_r)$ automatically satisfies (2.67)).

If ind < 0 then dim coker > 0. Hence, using (2.81) we can argue exactly as above to show that if ind < 0, then $\check{P}^*_{(x,\xi)}(\underline{s}, D_{\underline{s}}, D_r)$ automatically fails to satisfy (2.67), regardless of what $\sigma_{sub}(P^*)|_{(x,\xi)}$ is, and $\check{P}_{(x,\xi)}(\underline{s}, D_{\underline{s}}, D_r)$ automatically satisfies (2.67). These facts correspond to the results in the "abstract set-up" mentioned earlier. We mention also, in this connection, that if we were to treat the analogues of the first order factors $\partial_t - a_o'(0)tA$, $\partial_t - b_o'(0)tA$ from the viewpoint of the index, we would see, under the assumption that Re $a_o'(0)$ and Re $b_o'(0)$ have opposite sign, that one of the factors has index 1 and the other index -1. Since the index of a composition is the sum of the indices the index of $(\partial_t - a_o^1(0)tA)(\partial_t - b_o^1(0)tA)$ is 0. Thus the influences certainly have conflicted! The same occurs if we replace t by t^k with k any <u>odd</u> integer. This should be compared with (1.1), §1.

To complete the proofs of the above assertions concerning the cases ind > 0 and ind < 0, observe that, since $\tilde{P}_{(x,\xi)} + \sigma_{sub}(P)|_{(x,\xi)}$ and $\widetilde{P^*}_{(x,\xi)} + \sigma_{sub}(P^*)|_{(x,\xi)}$ are second-order linear ordinary differential operators, it follows, for example from Prop 2.22, that

(2.90) $\dim \ker(\tilde{P}_{(x,\xi)} + \sigma_{sub}(P)|_{(x,\xi)}) \leq 2$ and
$\dim \ker(\widetilde{P^*}_{(x,\xi)} + \sigma_{sub}(P^*)|_{(x,\xi)}) \leq 2$.

Hence, from Prop 2.32 and and (2.81) we see that

(2.91) If $n = 1$ and $\operatorname{ind}(\tilde{P}_{(x,\xi)} + \sigma_{sub}(P)|_{(x,\xi)}) > 0$
then $\dim \ker(\tilde{P}_{(x,\xi)} + \sigma_{sub}(P)|_{(x,\xi)}) = 2$ and
$\dim \ker(\widetilde{P^*}_{(x,\xi)} + \sigma_{sub}(P^*)|_{(x,\xi)}) = 0$.

(2.92) If $n = 1$ and $\operatorname{ind}(\tilde{P}_{(x,\xi)} + \sigma_{sub}(P)|_{(x,\xi)}) < 0$ then
$\dim \ker(\tilde{P}_{(x,\xi)} + \sigma_{sub}(P)|_{(x,\xi)}) = 0$ and
$\dim \ker(\widetilde{P^*}_{(x,\xi)} + \sigma_{sub}(P^*)|_{(x,\xi)}) = 2$

This completes the proof of the above assertions.

We shall show next how the sign of the index determines the nature of the set of eigenvalues of $\tilde{P}_{(x,\xi)} + \sigma_{sub}(P)|_{(x,\xi)}$. Since the index depends only on $\tilde{P}_{(x,\xi)}$ and not on the lower order part we see that the following stronger versions of (2.91) and (2.92) hold:

(2.91) If $n = 1$ and $\operatorname{ind}(\tilde{P}_{(x,\xi)} + \sigma_{sub}(P)|_{(x,\xi)}) > 0$,
then for every $\lambda \in \mathbb{C}$ $\dim \ker(\tilde{P}_{(x,\xi)} + \sigma_{sub}(P)|_{(x,\xi)} - \lambda) = 2$ and

$$\dim \ker (P^*_{(x,\xi)} + \sigma_{sub}(P^*)|_{(x,\xi)} - \lambda) = 0 .$$

(2.92) If $n = 1$ and $\mathrm{ind}\,(\widetilde{P}_{(x,\xi)} + \sigma_{sub}(P)|_{(x,\xi)}) < 0$, then for every $\lambda \in ¢$ $\dim \ker(\widetilde{P}_{(x,\xi)} + \sigma_{sub}(P)|_{(x,\xi)} - \lambda) = 0$ and $\dim \ker(\widetilde{P^*}_{(x,\xi)} + \sigma_{sub}(P^*)|_{(x,\xi)} - \lambda) = 2$.

In particular we see that if $\mathrm{ind} > 0$ then every $\lambda \in ¢$ is an eigenvalue of $\widetilde{P}_{(x,\xi)} + \sigma_{sub}(P)|_{(x,\xi)}$, and that if $\mathrm{ind} < 0$ then no $\lambda \in ¢$ is an eigenvalue of $\widetilde{P}_{(x,\xi)} + \sigma_{sub}(P)|_{(x,\xi)}$. In the latter case it is true, however, that every $\lambda \in ¢$ lies in the spectrum of $\widetilde{P}_{(x,\xi)} + \sigma_{sub}(P)|_{(x,\xi)}$ viewed as an unbounded Hilbert space operator, since $\dim \mathrm{coker}\,(\widetilde{P}_{(x,\xi)} + \sigma_{sub}(P)|_{(x,\xi)} - \lambda) > 0$.

We shall see when, in the case $\mathrm{ind} = 0$, we compute the eigenvalues of $\widetilde{P}_{(x,\xi)} + \sigma_{sub}(P)|_{(x,\xi)}$ that they form a countable set without limit points. In fact, this implies that the spectrum of $\widetilde{P}_{(x,\xi)} + \sigma_{sub}(P)|_{(x,\xi)}$ is discrete, for we have

Lemma 2.33. If $n = 1$ and $\mathrm{ind}(\widetilde{P}_{(x,\xi)} + \sigma_{sub}(P)|_{(x,\xi)}) = 0$, then if $\lambda \in ¢$ lies in the spectrum of $\widetilde{P}_{(x,\xi)} + \sigma_{sub}(P)|_{(x,\xi)}$ λ is actually an eigenvalue of $\widetilde{P}_{(x,\xi)} + \sigma_{sub}(P)|_{(x,\xi)}$.

<u>Pf</u>:

Let λ be an element of $\mathbb{C}$. Since ind depends only on $\widetilde{P}_{(x,\xi)}$ we get

$$(2.93) \quad \operatorname{ind}(\widetilde{P}_{(x,\xi)} + \sigma_{sub}(P)|_{(x,\xi)} - \lambda) = 0 .$$

Since $P^*_{(x,\xi)} + \sigma_{sub}(P^*)|_{(x,\xi)} - \bar{\lambda} =$

$(\widetilde{P}_{(x,\xi)} + \sigma_{sub}(P)|_{(x,\xi)} - \lambda)^*$ it follows from (2.93) that

$$(2.94) \quad \operatorname{ind}(\widetilde{P^*}_{(x,\xi)} + \sigma_{sub}(P^*)|_{(x,\xi)} - \bar{\lambda}) = 0 .$$

If λ is not an eigenvalue of $\widetilde{P}_{(x,\xi)} + \sigma_{sub}(P)|_{(x,\xi)}$ it follows from (2.93) that $\dim\ker(\widetilde{P^*}_{(x,\xi)} + \sigma_{sub}(P^*)|_{(x,\xi)} - \bar{\lambda}) = 0$. This together with (2.94) $\Longrightarrow$

$$(2.95) \quad \operatorname{range}(\widetilde{P^*}_{(x,\xi)} + \sigma_{sub}(P^*)|_{(x,\xi)} - \bar{\lambda}) = \text{all of } L^2 .$$

It follows essentially from the closed range theorem (see, [28] Cor 1., p.208) that $\widetilde{P} + \sigma_{sub}(P)|_{(x,\xi)} - \lambda$ has a continuous inverse.

<u>QED</u>

We conclude with the remark that if $n = 1$ and p is real-valued, then the roots of (2.78) obviously have opposite sign, so ind = 0 .

§2.5. Computation of eigenvalues

Our next order of business will be to actually compute the eigenvalues of $\tilde{P}_{(x,\xi)}$. We shall do this in the general case of p complex-valued by a generalized version of the physicist's procedure (see [21] Chapter XII) for computing the eigenvalues of the harmonic oscillator. This is, essentially, Treves' method of concatenations. We shall see that, in the case of real p, $\tilde{P}_{(x,\xi)}$ consists of n independent harmonic oscillators with real "weights", and that in the general case $\tilde{P}_{(x,\xi)}$ at least has the same eigenvalues as an operator consisting of n independent harmonic oscillators with complex weights. This will "explain" the n-parameter family of discrete conditions appearing in [4]. We shall also discuss the eigenvalues from the viewpoint of Maslov asymptotics ([19], [40]).

We begin by providing an optimal normal form for $\tilde{P}_{(x,\xi)}$ in case p is real-valued; more precisely, we find an optimal representative in the unitary equivalence class of operators $\tilde{P}_{(x,\xi)}$. Our procedure will be a variant of methods in [4].

Recall (Lemma 2.8) that if p is real-valued, then for every $(x,\xi) \in \Sigma$, $\tilde{p}_{(x,\xi)}$ is strictly definite. Hence, given that p is real-valued, we assume loss of generality that $\tilde{p}_{(x_o,\xi_o)}$ is positive-definite. We want to show that for some choice of symplectic linear coordinates $s_1,\dots\dots,s_n$, $t_1,\dots\dots,t_n$, $\tilde{p}_{(x,\xi)}$ may be put into "diagonal form"

$$(2.96) \quad \tilde{p}_{(x,\xi)}(\underline{s}, \underline{t}) = \sum_{i=1}^{n} a_i(t_i^2 + s_i^2) \quad , \quad a_i > 0$$

It will then follow, in particular, that

$$(2.97) \quad \tilde{P}_{(x,\xi)} = \sum_{i=1}^{n} a_i \, (D_{s_i}^2 + s_i^2) \quad .$$

That is, $\tilde{P}_{(x,\xi)}$ will consist of n independent harmonic oscillators with "weights" a_i . These a_i (at each fixed point (x,ξ)) are, in fact, uniquely determined (up to permutation, of course).

Remark 2.34 By using the symplectic transformation

$$g: \; s_i \longmapsto \lambda_i s_i \quad , \; t_i \longmapsto \frac{1}{\lambda_i} t_i \quad \text{with } \lambda_i \text{ equal to } \sqrt{a_i}$$

we get that

$$(2.98) \quad g\,\tilde{p}_{(x,\xi)}(\underline{s}, \underline{t}) = \sum_{i=1}^{n} (a_i t_i)^2 + s_i^2$$

so that we could also take as representative for $\tilde{P}_{(x,\xi)}$

$$(2.99) \quad \tilde{P}_{(x,\xi)} = \Sigma \, (a_i \, D_{s_i})^2 + s_i^2 \quad .$$

Similarly we see that we could take as representative for $\widetilde{P}_{(x,\xi)}$

$$(2.100) \qquad \widetilde{P}_{(x,\xi)} = \Sigma\, D_{s_i}^{2} + a_i^{2} s_i^{2} \quad .$$

Hence, we can view the a_i's as frequencies rather than as "weights" if we prefer.

Let us now get on with the actual proof. The non-degenerate anti-symmetric bilinear form $\omega_{(x,\xi)}$ on $N(\Sigma)_{(x,\xi)}$ induces a Hermitian-symmetric form $\hat{\omega}_{(x,\xi)}$ on $N(\Sigma)_{(x,\xi)} \otimes \mathbb{C}$ defined by

$$(2.101) \qquad \hat{\omega}_{(x,\xi)}\,(\zeta_1, \zeta_2) \equiv \frac{1}{\sqrt{-1}}\ \omega_{(x,\xi)}\,(\zeta_1, \overline{\zeta}_2) \quad ,$$

where we denote by $\omega_{(x,\xi)}$ the natural extension of $\omega_{(x,\xi)}$ to the complexified space. Since $\omega_{(x,\xi)}$ is non-degenerate on $N(\Sigma)_{(x,\xi)}$ it follows that $\hat{\omega}_{(x,\xi)}$ is non-degenerate on $N(\Sigma)_{(x,\xi)} \otimes \mathbb{C}$. We also consider that form $\hat{p}_{(x,\xi)}$ on $N(\Sigma)_{(x,\xi)} \otimes \mathbb{C}$ defined by

$$(2.102) \qquad \hat{p}_{(x,\xi)}(\zeta_1, \zeta_2) = \widetilde{p}_{(x,\xi)}\,(\zeta_1, \overline{\zeta}_2) \quad ,$$

where we continue to denote by $\widetilde{p}_{(x,\xi)}$ the extension of $\widetilde{p}_{(x,\xi)}$ to the complexified space. Since $\widetilde{p}_{(x,\xi)}$ on $N(\Sigma)_{(x,\xi)}$ is assumed positive-definite, in particular <u>real</u>, it follows that $\hat{p}_{(x_0,\xi_0)}$ is in fact Hermitian symmetric, indeed, positive-definite.

We shall show how to construct a subspace W (over $\mathbb{C}$) of $N(\Sigma)_{(x,\xi)} \otimes \mathbb{C}$ satisfying the properties:

(2.103) $\dim_{\mathbb{C}} W = n = \frac{1}{2} \dim_{\mathbb{R}} N(\Sigma)_{(x,\xi)}$

(2.104) $\hat{\omega}_{(x,\xi)}$ is positive-definite on W and $\hat{p}_{(x,\xi)}$ is positive-definite on W. (Of course, as we pointed out earlier, $\hat{p}_{(x,\xi)}$ is positive-definite on $N(\Sigma)_{(x,\xi)} \otimes \mathbb{C}$, a fortiori on any complex subspace of $N(\Sigma)_{(x,\xi)} \otimes \mathbb{C}$.)

(2.105) W is orthogonal to $\overline{W}$, the conjugate subspace, both with respect to $\hat{p}_{(x,\xi)}$ and with respect to $\hat{}_{(x,\xi)}$ (and so, by (12.103) $N(\Sigma)_{(x,\xi)} \otimes \mathbb{C} = W \oplus \overline{W}$.)

First, notice that since $\hat{\omega}_{(x,\xi)}$ is non-degenerate there exists a unique linear map $A : N(\Sigma)_{(x,\xi)} \otimes \mathbb{C} \longrightarrow N(\Sigma)_{(x,\xi)} \otimes \mathbb{C}$ defined by

(2.106) $\hat{\omega}_{(x,\xi)}(Au, v) = \hat{p}_{(x,\xi)}(u, v)$ for every $u, v \in N(\Sigma)_{(x,\xi)} \otimes \mathbb{C}$.

We shall find it convenient, under the present assumption that p is real valued, to work with A^{-1} rather than A .

<u>Lemma 2.35</u>. A does not have 0 as an eigenvalue. In particular, A^{-1} exists.

<u>Pf:</u>

Suppose $Au = 0$. Then, by (2.106) , $\hat{p}_{(x,\xi)}(u, v) = 0$ for every $v \in N(\Sigma)_{(x,\xi)} \otimes \mathbb{C}$. Since $\hat{p}_{(x,\xi)}$ is non-degenerate it follows that $u = 0$.

QED

<u>Lemma 2.36.</u> A^{-1} is symmetric with respect to $\hat{p}_{(x,\xi)}$. Therefore, since $\hat{p}_{(x,\xi)}$ is positive-definite Hermitian, it follows that all the eigenvalues of A^{-1} are real, and that $N(\Sigma)_{(x,\xi)} \otimes \mathbb{C}$ is the direct sum of genuine (not just generalized) eigenspaces of A^{-1} , <u>orthogonal</u> with respect to $\hat{p}_{(x,\xi)}$.

<u>Pf:</u>

$$\hat{p}_{(x,\xi)}(A^{-1} u, v) = \hat{\omega}_{(x,\xi)} (u, v) = \overline{\hat{\omega}_{(x,\xi)} (v, u)}$$

$$= \overline{\hat{p}_{(x,\xi)}(A^{-1}v, u)} = \hat{p}_{(x,\xi)} (u, A^{-1}v) .$$

QED

<u>Definition 2.37.</u> Let W be the subspace of $N(\Sigma)_{(x,\xi)} \otimes \mathbb{C}$ (over $\mathbb{C}$) spanned by the eigenvectors of A^{-1} with <u>positive</u>

eigenvalues.

We shall show that W satisfies (2.103) - (2.105) .

Lemma 2.38. $\hat{\omega}_{(x,\xi)}$ is positive definite on W .

Pf:

Let $u \in W$. Then, by Lemma 2.36, $u = u_1 + \cdots + u_\ell$ where $u_i \perp u_j$ with respect to $\hat{p}_{(x,\xi)}$ for $i \neq j$, and where $A^{-1} u_i = \lambda_i u_i$, $\lambda_i > 0$. Therefore, $\hat{\omega}_{(x,\xi)}(u, u) = \Sigma \lambda_i \hat{p}_{(x,\xi)}(u_i, u_i)$. Since $\hat{p}_{(x,\xi)}$ is positive definite this number is > 0 unless $u = 0$.

QED

Lemma 2.39. $\dim_{\mathbb{C}} W = n$.

Pf:

In view of Lemmas 2.35 and 2.36 it suffices to show that if u is an eigenvector of A^{-1} with $A^{-1}u = \lambda u$, then $\bar{u}$ is an eigenvector of A^{-1} with $A^{-1}\bar{u} = -\lambda u$ for this shows that if W^- is the subspace of $N(\Sigma)_{(x,\xi)} \otimes \mathbb{C}$ spanned by the eigenvectors of A^{-1} with negative eigenvalues then $\dim_{\mathbb{C}} W = \dim_{\mathbb{C}} W^-$, and Lemmas 2.35 and 2.36 imply that $\dim_{\mathbb{C}} (N(\Sigma)_{(x,\xi)} \otimes \mathbb{C}) = \dim_{\mathbb{C}} W + \dim_{\mathbb{C}} W^-$.

We now prove the above assertion by referring to the definition (2.101) and (2.102) . Indeed, if

$$\hat{\omega}_{(x,\xi)}(u, v) = \lambda\, \hat{p}_{(x,\xi)}(u, v) \quad \text{for every} \quad v \in N(\Sigma)_{(x,\xi)} \otimes ¢$$

$$\frac{1}{\sqrt{-1}}\, \omega_{(x_o,\xi_o)}(u, \bar{v}) = \lambda\, \tilde{p}_{(x,\xi)}(u, \bar{v}) \quad \text{, and so}$$

$$\overline{\frac{1}{\sqrt{-1}}\, \omega_{(x,\xi)}(u, \bar{v})} = \lambda\, \overline{\tilde{p}_{(x,\xi)}(u, \bar{v})} \quad (\text{since } \lambda \text{ is real}).$$

But, since $\omega_{(x,\xi)}$ and $\tilde{p}_{(x,\xi)}$ are both real on $N(\Sigma)$, this means that

$$-\frac{1}{\sqrt{-1}}\, \omega_{(x,\xi)}(\bar{u}, v) = \lambda\, \tilde{p}_{(x,\xi)}(\bar{u}, v) .$$

That is, $\frac{1}{\sqrt{-1}}\, \omega_{(x,\xi)}(\bar{u}, v) = -\lambda\, \tilde{p}_{(x,\xi)}(\bar{u}, v)$ for every $v \in N(\Sigma)_{(x,\xi)} \otimes ¢$.

QED

Lemma 2.40. W is orthogonal to $\overline{W}$, the conjugate subspace, both with respect to $\hat{p}_{(x,\xi)}$ and with respect to $\hat{\omega}_{(x,\xi)}$.

Pf:

We showed in the proof of Lemma 2.39 that $\overline{W} = W^-$. Thus, it suffices to show that if $A^{-1}u = \lambda_1 u$ with $\lambda_1 > 0$ and $A^{-1}v = \lambda_2 v$ with $\lambda_2 < 0$, then $\hat{\omega}_{(x,\xi)}(u, v) = 0$

and $\hat{p}_{(x,\xi)}(u, v) = 0$. But

$$\lambda_1 \hat{p}_{(x,\xi)}(u, v) = \hat{p}_{(x,\xi)}(A^{-1}u, v) = \hat{\omega}_{(x,\xi)}(u, v),$$

and, since λ_2 is real,

$$\lambda_2 \hat{p}_{(x,\xi)}(u, v) = \hat{p}_{(x,\xi)}(u, A^{-1}v) = \hat{p}_{(x,\xi)}(A^{-1}u, v) = \hat{\omega}_{(x,\xi)}(u, v) .$$

Since $\lambda_1 \neq \lambda_2$ it follows that $\hat{p}_{(x,\xi)}(u, v) = 0$, and , therefore, also that $\hat{\omega}_{(x,\xi)}(u, v) = 0$.

QED

Thus, we have constructed a subspace W satisfying (2.103) - (2.105) . By (2.104) we have two positive-definite Hermitian inner products on the complex vector space W . Thus, by the standard linear algebra result, there is a basis $\zeta_1, \ldots, \zeta_n$ for W orthonormal with respect to $\hat{\omega}_{(x,\xi)}$ and orthogonal with respect to $\hat{p}_{(x,\xi)}$, with positive eigenvalues.

That is,

(2.107) $\hat{\omega}_{(x,\xi)}(\zeta_i, \zeta_j) = \delta_{ij}$

(2.108) $\hat{p}_{(x,\xi)}(\zeta_i, \zeta_j) = a_i \delta_{ij}$, $a_i > 0$.

Of course, the a_i 's are just the positive eigenvalues of A .

But (2.107) can be rewritten as

$$(2.109) \quad \frac{1}{\sqrt{-1}} \omega_{(x,\xi)} (\zeta_i, \overline{\zeta}_j) = \delta_{ij} \quad .$$

and (2.105) implies

$$(2.110) \quad \frac{1}{\sqrt{-1}} \omega_{(x,\xi)} (\zeta_i, \zeta_j) = 0 \quad .$$

If we write $\zeta_i = \dfrac{u_i + \sqrt{-1}\, v_i}{\sqrt{2}}$, $u_i, v_i \in N(\Sigma)_{(x,\xi)}$,

then (2.109) and (2.110) together simply say that $u_1, \ldots, u_n$, $v_1, \ldots, v_n$ form a symplectic basis for $N(\Sigma)_{(x,\xi)}$, i.e.,

$$(2.111) \quad w_{(x,\xi)}(u_i, u_j) = 0 \ , \ w_{(x,\xi)}(v_i, v_j) = 0 \ ,$$
$$w_{(x,\xi)}(u_i, v_j) = -\delta_{ij}$$

Rewriting (2.108), we have

$$(2.112) \quad \widetilde{p}_{(x,\xi)}(\zeta_i, \overline{\zeta}_j) = a_i \ \delta_{ij}$$

Again using (2.105) we get, in addition,

$$(2.113) \quad \widetilde{p}_{(x,\xi)}(\zeta_i, \zeta_j) = 0 \ .$$

(2.112) and (2.113) together simply say that

$$(2.114) \quad \widetilde{p}_{(x,\xi)}(u_i, u_j) = a_i \ \delta_{ij} \ , \ \widetilde{p}_{(x,\xi)} (v_i, v_j) = a_i \ \delta_{ij} \ ,$$
$$\widetilde{p}_{(x,\xi)}(u_i, v_j) = 0 \ .$$

If we use the linear coordinate system $(\underline{s}, \underline{t}) \longmapsto \Sigma s_i u_i + \Sigma t_i v_i$ then (2.111) says that $s_1, \ldots, s_n, t_1, \ldots, t_n$ are symplectic linear coordinates on $N(\Sigma)_{(x,\xi)}$, and (2.114) says that $\tilde{p}_{(x,\xi)}$, expressed in these coordinates has the form $\tilde{p}_{(x,\xi)}(\underline{s}, \underline{t}) = \sum_{i=1}^{n} a_i (t_i^2 + s_i^2)$, i.e., the form (2.96).

We show next that the a_i's occurring in (2.96) are invariantly determined by $\tilde{p}_{(x,\xi)}$ and $\omega_{(x,\xi)}$. That is, for any choice of symplectic (with respect to $\omega_{(x,\xi)}$) linear coordinates $\underline{s}, \underline{t}$ in which $\tilde{p}_{(x,\xi)}$ can be written as $\tilde{p}_{(x,\xi)}(\underline{s}, \underline{t}) = \sum_{i=1}^{n} b_i (t_i^2 + s_i^2)$, the b_i's (counted with multiplicities) are the same as the a_i's. Indeed, suppose that in the symplectic linear coordinates $\underline{s}, \underline{t}$, corresponding to a symplectic basis $u_1', \ldots, u_n', v_1', \ldots v_n'$, $\tilde{p}_{(x,\xi)}(\underline{s}, \underline{t}) = \sum_{i=1}^{n} b_i (t_i^2 + s_i^2)$. Then $\omega_{(x,\xi)}(u_i', u_j') = 0$, $\omega_{(x,\xi)}(v_i', v_j') = 0$, and $\omega_{(x,\xi)}(u_i', v_j') = -\delta_{ij}$; also, $\tilde{p}_{(x,\xi)}(u_i', u_j') = b_i \delta_{ij}$, $\tilde{p}_{(x,\xi)}(v_i', v_j') = b_i \delta_{ij}$, and $\tilde{p}_{(x,\xi)}(u_i', v_j') = 0$. Reversing the steps of the preceding derivations we see that if

$$\zeta_i' \equiv \frac{u_i' + \sqrt{-1}\, v_i'}{\sqrt{2}}, \text{ then } \zeta_1', \ldots, \zeta_n', \bar{\zeta}_1', \ldots, \bar{\zeta}_n'$$

form a basis for $N(\Sigma)_{(x,\xi)} \otimes \mathbb{C}$ and

$$\hat{\omega}_{(x,\xi)}(\zeta_i{}', \zeta_j{}') = \delta_{ij}, \quad \hat{\omega}_{(x,\xi)}(\zeta_i{}', \bar{\zeta}_j{}') = 0$$

and

$$\hat{p}_{(x,\xi)}(\zeta_i{}', \zeta_j{}') = b_i\, \delta_{ij}, \quad \hat{p}_{(x,\xi)}(\zeta_i{}', \bar{\zeta}_j{}') = 0$$

Hence, from (2.106) it follows that $A\zeta_i{}' = b_i\, \zeta_i{}'$ (and, thus, also that $A^{-1}\zeta_i{}' = b_i{}^{-1}\, \zeta_i{}'$).
Since $\tilde{p}_{(x_0,\xi_0)}$ is positive definite the b_i 's are > 0. Hence, the $\zeta_i{}'$'s are a basis for the n-dimensional subspace W of Def. 2.37, and so the b_i's are the positive eigenvalues of A. Thus the b_i 's are the same as the a_i 's. We may summarize the preceding in the following proposition:

<u>Proposition 2.41.</u> If $\tilde{p}_{(x_0,\xi_0)}$ is positive-definite, then there exist symplectic (with respect to $\omega_{(x,\xi)}$) linear coordinates $s_1, s_2, \ldots, s_n$, $t_1, \ldots, t_n$ for $N(\Sigma)_{(x,\xi)}$ with respect to which $\tilde{p}_{(x,\xi)}$ has the "diagonal form"

$$\tilde{p}_{(x,\xi)}(\underline{s}, \underline{t}) = \sum_{i=1}^{n} a_i\, (t_i{}^2 + s_i{}^2).$$

These a_i 's are uniquely determined (up to permutation), being the same in any "diagonal form" representation of $\tilde{p}_{(x,\xi)}$ with respect to linear symplectic coordinates. Indeed, the a_i's are precisely the positive eigenvalues (counted with multiplicities) of the linear transformation $A : N(\Sigma)_{(x,\xi)} \otimes ¢ \longrightarrow N(\Sigma)_{(x,\xi)} \otimes ¢$ defined by (2.106).

As we noted earlier, if $\tilde{p}_{(x,\xi)}(\underline{s}, \underline{t}) = \sum_{i=1}^{n} a_i(t_i^2 + s_i^2)$, then we may take as a representative of the unitary equivalence class of operators $\tilde{P}_{(x,\xi)}$ the operator $\sum_{i=1}^{n} a_i (D_{s_i}^2 + s_i^2)$. Now it is well-known that the eigenvalues of the harmonic oscillator $D_s^2 + s^2$ are precisely all numbers of the form $2n + 1$ where n is a non-negative integer. Moreover, there exists an orthonormal basis in $L^2(s)$ of eigenfunctions. $H_o(s), H_1(s), \ldots,$ (in $\mathscr{S}$), namely the Hermite functions. It then follows that the functions of the form $H_{n_1}(s_1) H_{n_2}(s_2) \cdots H_{n_n}(s_n)$ form an orthonormal basis for $L^2(s_1, \ldots, s_n)$. That is, $L^2(s_1, \ldots, s_n)$ has an orthonormal basis of eigenfunctions of $\tilde{P}_{(x,\xi)}$ with eigenvalues all numbers of the form $\sum_{i=1}^{n} a_i(2n_i + 1)$ where the n_i 's are non-negative integers. Thus,

(2.115) If $\tilde{p}_{(x,\xi)}$ has the form (2.96), then the eigenvalues of $\tilde{P}_{(x,\xi)}$ are precisely all numbers of the form $\sum_{i=1}^{n} a_i (2n_i + 1)$ where the n_i 's are non-negative integers. (Proof: If $\tilde{P}_{(x,\xi)}f = \lambda f$, then letting $H_{\underline{n}} = H_{n_1}(s_1) \cdots H_{n_n}(s_n)$, $\sum_{i=1}^{n} (2n_i + 1) \langle f, H_{\underline{n}} \rangle = \langle f, \tilde{P}_{(x,\xi)} H_{\underline{n}} \rangle = \langle \tilde{P}_{(x,\xi)}f, H_{\underline{n}} \rangle = \lambda \langle f, H_{\underline{n}} \rangle$. Thus, if

λ does not equal $\sum_{i=1}^{n} a_i(2n_i + 1)$ for any $\underline{n}$, $\langle f , H_{\underline{n}} \rangle = 0$ for every $\underline{n}$, and so $f = 0$.)

In the general case of p complex-valued, $\tilde{p}_{(x, \ )}$ on $N(\Sigma)_{(x,\xi)}$ is also complex-valued. Hence, we cannot conclude that $\hat{p}_{(x,\xi)}$ on $N(\Sigma)_{(x,\xi)} \otimes ¢$ is Hermitian symmetric, and so we cannot proceed exactly as above. However, if we make the additional assumption in case $n = 1$ that we have "conflicting influences", then it is possible to construct a subspace W (over $¢$) of $N(\Sigma)_{(x,\xi)} \otimes ¢$ satisfying all the properties (2.103) - (2.105) <u>except</u> the property that $\hat{p}_{(x,\xi)}$ is positive-definite on W . In fact W can be taken as the vector subspace of $N(\Sigma)_{(x,\xi)} \otimes ¢$ spanned by all the generalized eigenvectors of A corresponding to eigenvalues a_i such that $\mathrm{Re}\_za_i) > 0$, where z is a certain complex number. (By a generalized eigenvector v corresponding to a_i we mean a non-zero vector v such that $(A - a_i)^{\ell} v = 0$ for some ℓ.) Since the proof may be found in [4] we shall not present it here in full. We shall, however, indicate now the "conflicting influences" assumption comes into the argument. Recall (Remark 2.9) that

(2.116) If $n > 1$ then there exists $z \in ¢$ such that $\mathrm{Re}\, z\, \tilde{p}_{(x,\xi)}$ is positive-definite. The same holds for $n = 1$ if we assume "conflicting influences".

We shall show, by using (2.116) , that zA has

precisely n eigenvalues with positive real part and n eigenvalues with negative real part, where A is defined by (2.106).

Pf:

Writing $z\,\tilde{p}_{(x,\xi)} \equiv \tilde{p}^{(1)}_{(x,\xi)} + \sqrt{-1}\,\tilde{p}^{(2)}_{(x,\xi)}$ we have by (2.116) that $\tilde{p}^{(1)}_{(x,\xi)}$ is positive-definite. Now letting $\tilde{p}_{(x,\xi)}$, $\tilde{p}^{(1)}_{(x,\xi)}$, $\tilde{p}^{(2)}_{(x,\xi)}$ also denote the natural extensions to the complexified space $N(\Sigma)_{(x,\xi)} \otimes ¢$ we define $\tilde{p}^{t}_{(x,\xi)}$ for $t \in [0, 1]$ by

$$(2.117) \quad z\,\tilde{p}^{t}_{(x,\xi)} \equiv \tilde{p}^{(1)}_{(x,\xi)} + \sqrt{-1}\; t\,\tilde{p}^{(2)}_{(x,\xi)}$$

and define $\hat{p}^{t}_{(x,\xi)}$ and A^t by

$$(2.118) \quad \hat{p}^{t}_{(x,\xi)}\,(\zeta_1, \zeta_2) = \tilde{p}^{t}_{(x,\xi)}(\zeta_1, \overline{\zeta_2}) \;.$$

$$(2.119) \quad \hat{\omega}_{(x,\xi)}\,(A^t \zeta_1, \zeta_2) = \hat{p}^{t}_{(x,\xi)}\,(\zeta_1, \zeta_2) \;.$$

Observe that $\hat{p}^{1}_{(x,\xi)} = \hat{p}_{(x,\xi)}$ (defined by (2.102)) and $A^1 = A$ (defined by (2.106)) . When $t = 0$, then since $\tilde{p}^{(1)}_{(x,\xi)}$ is positive-definite so is $z\,\hat{p}^{\circ}_{(x,\xi)}$, and so we are precisely in the case treated earlier. In particular, zA° has all its eigenvalues real, n strictly positive and n strictly negative. One checks easily that A^t varies continuously with t, and, in particular, that the eigenvalues of zA^t can be represented as continuous functions of t. It follows that to prove

(2.120) zA^1 has precisely n eigenvalues with positive real part and n eigenvalues with negative real part

it will suffice to prove that zA^t can never have a purely imaginary eigenvalue. But suppose that $zA^t\zeta = \lambda\zeta$ for some $\zeta \in N(\Sigma)_{(x,\xi)} \otimes ¢$ and some purely imaginary λ. Then by (2.119) and (2.117)

$$(2.121) \quad \lambda\, \hat{\omega}_{(x,\xi)}(\zeta, \zeta) = \tilde{p}^{(1)}_{(x,\xi)}(\zeta, \bar{\zeta}) + \sqrt{-1}\, t \, \sqrt{-1}\, t\ \tilde{p}^{(2)}_{(x,\xi)}(\zeta, \bar{\zeta}) \ .$$

Since $\tilde{p}^{(1)}_{(x,\xi)}$ is positive-definite on $N(\Sigma)_{(x,\xi)} \otimes ¢$ and since $\hat{\omega}_{(x,\xi)}(\zeta, \zeta)$ is real, (2.121) cannot hold unless $\zeta = 0$.

<u>QED</u>

Since z in (2.116) is **far from** unique, we would like a more "intrinsic" description of the eigenvalues a_i for which $\mathrm{Re}\,(za_i) > 0$. This is given by the following lemma.

Lemma 2.42. If $n > 1$ or if $n = 1$ and we assume "conflicting influences" then $\hat{\omega}_{(x,\xi)}$ is positive definite on the n-dimensional subspace W^+ of $N(\Sigma)_{(x,\xi)} \otimes ¢$ spanned by the generalized eigenvectors corresponding to n of the eigenvalues of A (counting multiplicities) and is negative definite on the n-dimensional subspace W^- spanned by the generalized eigenvectors corresponding to the remaining n eigenvalues of A. For any z satisfying (2.116) the eigenvalues a_i of A for which $\operatorname{Re}(za_i) > 0$ are precisely those associated with W^+. That is, the space W defined as the span of the generalized eigenvectors corresponding to these a_i 's equals W^+.

Pf: (adapted from [4])

We maintain the notation of the preceding proof. Let W^t be the subspace of $N(\Sigma)_{(x,\xi)} \otimes ¢$ spanned by the generalized eigenvectors corresponding to the eigenvalues $a_i{}^t$ of A^t for which $\operatorname{Re}(za_i{}^t) > 0$. We know from the preceding proof that $\dim_{¢} W^t = n$. We know, since the case $t = 0$ corresponds to $\tilde{p}_{(x,\xi)}$ positive-definite which we treated in detail earlier, that $\tilde{\omega}_{(x,\xi)}$ restricted to W^o is positive-definite. Let e_+, e_-, e_o be the functions from the space of Hermitian forms on $¢^n$ into the non-negative integers assigning to each form the dimension of a maximal positive-definite subspace, the dimension of a maximal negative-definite subspace, and the dimension of its null-space, respectively. Note that $n = e_+ + e_- + e_o$ and that both e_+ and e_- are lower semicontinuous, i.e., can only change by "jumping up". One can check that W^t varies continuously with t, and hence that $\hat{\omega}_{(x,\xi)}$ restricted to W^t varies continuously with t. It follows that if e_o does not vary with t then neither do e_+ or e_-. That is, if we can show that $\hat{\omega}_{(x,\xi)}$ restricted to W^t is non-degenerate for each t, then $\hat{\omega}_{(x,\xi)}$ restricted to W^t is positive-definite for every t, in particular for $t = 1$. We shall assume the fact (proved in [4]) that $\overline{W}^t$ is orthogonal to W^t with respect to $\hat{\omega}_{(x,\xi)}$. (This is part of (2.105)). Since $\dim_{\mathbb{C}} W^t = n = \frac{1}{2} \dim_{\mathbb{C}} N(\Sigma)_{(x,\xi)} \otimes \mathbb{C}$ it follows that if v is in the null-space of $\hat{\omega}_{(x,\xi)}$ restricted to W^t then v is in the null-space of $\hat{\omega}_{(x,\xi)}$ viewed as a Hermitian form on all of $N(\Sigma)_{(x,\xi)} \otimes \mathbb{C}$. Since $\hat{\omega}_{(x,\xi)}$ so viewed is non-degenerate it follows that $v = 0$. This proves that $\hat{\omega}_{(x,\xi)}$ is positive-definite on W. In exactly the same way one proves that $\hat{\omega}_{(x,\xi)}$ is negative-definite on the span of the generalized eigenvectors corresponding to the eigenvalues a_i such that $\mathrm{Re}(za_i) < 0$.

<u>QED</u>

We shall also make use of (2.116) in the following form.

Lemma 2.42A. If $\operatorname{Re} z\, \tilde{p}_{(x,\xi)}$ is positive-definite then the set $\{\operatorname{Re} z\lambda \mid \lambda$ is an eigenvalue of $\tilde{P}_{(x,\xi)}\}$ is bounded below by 0. (For our purposes any lower bound, even negative, would do just as well.)

Pf:

By Prop. 2.22 any eigenfunction v of $\tilde{P}_{(x,\xi)}$ lies in $\mathscr{S}$. Hence, letting $\langle , \rangle$ denote the L^2 inner product, it suffices to prove

(2.121a) $\operatorname{Re} \langle z\, \tilde{P}_{(x,\xi)} v, v \rangle \geqq 0$ for every $v \in \mathscr{S}$.

We write $z\, \tilde{p}_{(x,\xi)} \equiv \tilde{p}^{(1)}_{(x,\xi)} + \sqrt{-1}\; \tilde{p}^{(2)}_{(x,\xi)}$ where $\tilde{p}^{(1)}_{(x,\xi)}$ and $\tilde{p}^{(2)}_{(x,\xi)}$ are real, and where, in fact, $\tilde{p}^{(1)}_{(x,\xi)}$ is positive-definite, by hypothesis. Passing to the test-operators we get

$$(2.121b)\quad z\, \tilde{P}_{(x,\xi)} = \tilde{P}^{(1)}_{(x,\xi)} + \sqrt{-1}\; \tilde{P}^{(2)}_{(x,\xi)} .$$

Since $\tilde{p}^{(1)}_{(x,\xi)}$ is positive-definite it follows from Prop. 2.41 that there are positive numbers $a_1, \dots, a_n$ such that for every $v \in \mathscr{S}$

$$(2.121c)\quad \langle \tilde{P}^{(1)}_{(x,\xi)} v, v \rangle = \sum_{i=1}^{n} a_i \langle (D_{s_i}^2 + s_i^{\,2})\, v, v \rangle$$

$$= \sum_{i=1}^{n} a_i \left(\langle D_{s_i} v, D_{s_i} v \rangle + \langle s_i v, s_i v \rangle \right) \geqq 0 .$$

On the other hand, recalling the definition of the test-operators, we see that since $\tilde{p}^{(2)}_{(x,\xi)}$ is real $\tilde{P}^{(2)}_{(x,\xi)}$ is a finite linear combination with real coefficients of formally self-adjoint differential operators.

It follows that

$$(2.121d) \quad \langle \tilde{P}^{(2)}_{(x,\xi)} v, v \rangle \text{ is real} \qquad \text{for every } v \in \mathscr{S}$$

(2.121c) and (2.121d) yield (2.121a) .

QED

Let us now draw the consequences in the case $n = 1$ (with conflicting influences) which follow from the existence of W satisfying the properties (2.103) - (2.105) (with the exception of the property that $\hat{p}_{(x,\xi)}$ is positive-definite on W). Just as before we can choose a basis ζ for W such that (2.109) and (2.110) hold. (Of course, since $n = 1$, i,j only take on the value 1.) Hence, if we define $u,v \in N(\Sigma)_{(x,\xi)}$ by

$$(2.122) \qquad \zeta = \frac{u + \sqrt{-1}\, v}{\sqrt{2}}$$

then u,v form a symplectic basis for $N(\Sigma)_{(x,\xi)}$, i.e.,

(2.111) holds. Let

$$(2.123) \quad a \equiv \hat{p}_{(x,\xi)}(\zeta, \zeta) = \tilde{p}_{(x,\xi)}(\zeta, \bar{\zeta}) \ .$$

Using (2.105) we get that

$$\tilde{p}_{(x,\xi)}(\zeta, \zeta) = 0 \ .$$

Hence we again get (2.114) :

$$\tilde{p}_{(x,\xi)}(u, u) = a \ , \ \tilde{p}_{(x,\xi)}(v, v) = a \ , \ \tilde{p}_{(x,\xi)}(u, v) = 0 \ .$$

Since (2.111) holds we may again define symplectic linear coordinates s,t for $N(\Sigma)_{(x,\xi)}$ by $(s, t) \longmapsto su + tv$. As before,

$$(2.124) \quad \tilde{p}_{(x,\xi)} = a(t^2 + s^2)$$

Recalling Lemma 2.42 we thus have the proposition

<u>Proposition 2.43.</u> If $n = 1$ and "conflicting-influences" hold, tne $\tilde{P}_{(x,\xi)} = a(D_s{}^2 + s^2)$, and so the eigenvalues of $\tilde{P}_{(x,\xi)}$ are precisely all numbers of the form $a\,(2n + 1)$ where n is a non-negative integer. Here a is the eigenvalue of A on whose eigenspace $\hat{\omega}_{(x,\xi)}$ is positive-definite.

We next treat the same case as above, namely $n = 1$ with "conflicting-influences" without first reducing to the "diagonal-form" (2.124). We do this as an illustration of the method of "concatenations". This will be the prototype of the computation we shall do later for the case $n > 1$, and also for the abstract case which we shall treat in §4.

Recall (see Prop. 2.31 and the subsequent discussion) that if $\tilde{p}_{(x,\xi)}(\underline{s}, \underline{t}) = a|_{(x,\xi)} t^2 + 2c|_{(x,\xi)} st + b|_{(x,\xi)} s^2$, then "conflicting influences" corresponds to the assumption that the equation

$$(2.125) \quad a|_{(x,\xi)} \zeta^2 + 2c|_{(x,\xi)} \zeta + b|_{(x,\xi)} = 0$$

has two roots α, β with imaginary parts respectively < 0 and > 0 . ($a|_{(x,\xi)}$ should not be confused with a appearing in (2.124), where a different set of symplectic coordinates is being used.) Since α,β are the roots of equation (2.125) it follows that

$$(2.126) \quad \tilde{p}_{(x,\xi)}(\underline{s}, \underline{t}) = a|_{(x,\xi)} (t - \alpha s)(t - \beta s)$$

Remembering that $\tilde{P}_{(x,\xi)}$ (more precisely, a representation of the unitary equivalence class $\tilde{P}_{(x,\xi)}$) is the differential operator having $\tilde{p}_{(x,\xi)}$ as its "symbol" and having no lower order terms in its "symmetric representation", we see that

$$(2.127) \quad \widetilde{P}_{(x,\xi)} = a|_{(x,\xi)} \left[D_s^{\,2} - \frac{\alpha+\beta}{2}(sD_s + D_s s) + \alpha\beta\, s^2\right]$$

$$= a|_{(x,\xi)}\left[(D_s - \alpha s)(D_s - \beta s)\right] + a|_{(x,\xi)}\left(\frac{\beta-\alpha}{2i}\right)$$

In particular, it is clear that we know the eigenvalues of $\widetilde{P}_{(x,\xi)}$ once we determine the eigenvalues of $N \equiv (D_s - \alpha s)(D_s - \beta s)$. We shall do this below. The reader may wish to compare this with the computation in ([21], Chapter XII) of the eigenvalues of the harmonic oscillator.

Let $X = D_s - \alpha s$ and $Y = D_s - \beta s$.

Since $\operatorname{Im} \alpha < 0$ and $\operatorname{Im} \beta > 0$ it is easy to check by a direct computation that

(2.128) X is injective on $\mathscr{S}$

and

(2.129) Y is not injective on $\mathscr{S}$

(Of course, it is possible to show more, for example, that as operators from $H_{(1,1)}(\mathbb{R}^1)$ to $L^2(\mathbb{R}^1)$ X is injective but not surjective and vice-versa for Y. However, since by Prop. 2.22 any eigenfunction of $\widetilde{P}_{(x,\xi)}$, and hence of N lies in $\mathscr{S}$, (2.128) and (2.129) will be sufficient for our purposes.)

Let $\delta \equiv \frac{1}{i}(\beta - \alpha) = [Y, X]$. Since $N = XY$ it follows from the definition of δ that

(2.130) $NX = X(N - \delta)$

(2.131) $NY = Y(N - \delta)$.

We work first with (2.131) . (2.131) shows that if v is an eigenvector of N with $Nv = \lambda v$, then Yv satisfies: $N(Yv) = (\lambda - \delta)Yv$. Notice also that if $Yv = 0$ then $Nv \equiv XYv = 0$, and so, since $v \neq 0$, $\lambda = 0$. Thus, if $\lambda \neq 0$, Yv is an eigenvector of N with eigenvalue $\lambda - \delta$. Iterating this process we see that

(2.132) For every positive integer n , if $Y^{n-1} v \neq 0$ and if $\lambda -(n - 1)\delta \neq 0$, then $Y^n v$ is an eigenvector of N , with eigenvalue $\lambda - n\delta$.

A simple computation, which we omit, shows that, under our assumption that $\operatorname{Im} \alpha < 0$ and $\operatorname{Im} \beta > 0$, we have

(2.133a) $\operatorname{Re}\big(\overline{\delta}\,(t - \alpha s)(t - \beta s)\big) \geqq 0$ for every $s,t \in \mathbb{R}$.

In view of the definition of N it follows from (2.126) , (2.127), and Lemma 2.42A that

(2.133b) The set $\{\mathrm{Re}\,\bar{\delta}\lambda \mid \lambda$ is an eigenvalue of $N\}$ is bounded below by a real number C $\left(= -\frac{|\delta|^2}{2}\right)$. (The property (2.133b) replaces the immediately verifiable fact in the case $N = Y^*Y$ of the harmonic oscillator that all the eigenvalues of N are non-negative.)

Since $\bar{\delta}\delta > 0$ it follows from (2.132) and (2.133b) that if n is a sufficiently large positive integer then $Y^n v = 0$. Choose the smallest such n. Then, again by (2.132), $\lambda - (n-1)\delta = 0$. Thus, we have shown

(2.134) If λ is an eigenvalue of N, then $\lambda = m\delta$ for some non-negative integer m.

We show next that

(2.135) If m is a non-negative integer then $m\delta$ is an eigenvalue of N.

First observe that this is true for $m = 0$, i.e., 0 is an eigenvalue of N. Indeed, by (2.129) there exists $v \neq 0$ such that $Yv = 0$, and so $Nv \equiv XYv = 0$. Next we show that if m is an eigenvalue of N then so is $m + 1$. This follows from (2.130) and (2.128), for if $v \neq 0$ satisfies $Nv = m\delta v$, then by (2.130) $N(Xv) = (m+1)\,\delta\,(Xv)$, and by (2.128) $Xv \neq 0$.

Thus we see, under the "conflicting influences"

assumption, that the eigenvalues of N are precisely all numbers of the form $m\delta$ where m is a non-negative integer.

We next treat the general case of $n \geqq 1$ and p complex-valued. Although it does not seem possible in general to express $\widetilde{P}_{(x,\xi)}$ as a sum of harmonic oscillators, nevertheless $\widetilde{P}_{(x,\xi)}$ has the same eigenvalues as a sum of harmonic oscillators. Indeed we shall show

<u>Proposition 2.44.</u> If $n = 1$ and we assume "conflicting influences" or if $n > 1$, then the eigenvalues of $\widetilde{P}_{(x,\xi)}$ are all numbers of the form $\sum_{i=1}^{n} (2n_i + 1)$, where the n_i are arbitrary non-negative integers and where the a_i are the eigenvalues of A on whose generalized eigenspaces $\hat{\omega}_{(x,\xi)}$ is positive-definite.

This is the "general" form of Props. 2.41 and 2.43. The proof that we shall present below will again use Treves' method of concatenations (see, for example, [4] sections 4, 5 and 6).

We shall use the same notation as in Lemma 2.42. Without loss of generality we may assume that P has been replaced by zp, i.e., we may assume that $z = 1$, and that the a_i's associated with W have strictly positive real parts. Recall that $\dim_{\mathbb{C}} W = n$, $\hat{\omega}_{(x,\xi)}$ is positive-definite on W, and that W is orthogonal to $\overline{W}$, the conjugate subspace, both with respect to $\hat{p}_{(x,\xi)}$ and with respect to $\hat{\omega}_{(x,\xi)}$.

The first step in the proof of Prop. 2.44 is to express $\tilde{P}_{(x,\xi)}$ in a convenient form. Let $\zeta_1,\dots,\zeta_n$ be an orthonormal basis of W with respect to $\hat{\omega}_{(x,\xi)}$, i.e., (2.107) holds. Then, as before, defining $u_i, v_i \in N(\Sigma)_{(x,\xi)}$ by

$$(2.136) \qquad \zeta_i = \frac{u_i + \sqrt{-1}\, v_i}{\sqrt{2}}$$

we see that $u_1,\dots,u_n, v_1,\dots,v_n$ form a symplectic basis for $N(\Sigma)_{(x,\xi)}$, i.e., (2.111) holds. Thus, if we use the linear coordinate system

$$(2.137) \qquad (\underline{s}, \underline{t}) \longmapsto \Sigma\, s_i u_i + \Sigma\, t_i v_i$$

then $s_1,\dots,s_n, t_1,\dots,t_n$ are symplectic linear coordinates on $N(\Sigma)_{(x,\xi)}$. A simple computation then shows that if we define complex linear coordinates $z_i, \bar{z}_i$ on $N(\Sigma)_{(x,\xi)} \otimes \mathbb{C}$ by

$$(2.138) \qquad z_i = s_i + \sqrt{-1}\, t_i \quad , \quad \bar{z}_i = s_i - \sqrt{-1}\, t_i$$

then if $v \in N(\Sigma)_{(x,\xi)}$ equals $\sum_i (s_i u_i + t_i v_i)$, then v may also be expressed as

$$(2.139) \qquad v = \frac{1}{\sqrt{2}} \sum_i (\bar{z}_i \zeta_i + z_i \bar{\zeta}_i)$$

Thus, $\tilde{p}_{(x,\xi)}(\underline{s}, \underline{t}) \equiv \tilde{p}_{(x,\xi)}(v, v)$

$$= \frac{1}{2}\Big[\sum_{i,j} \tilde{p}_{(x,\xi)}(\zeta_i, \zeta_j)\,\bar{z}_i\,\bar{z}_j + \sum_{i,j} \tilde{p}_{(x,\xi)}(\zeta_i, \bar{\zeta}_j)\,\bar{z}_i\,z_j + \sum_{i,j} \tilde{p}_{(x,\xi)}(\bar{\zeta}_j, \zeta_i)\,z_j\,\bar{z}_i + \sum_{i,j} \tilde{p}_{(x,\xi)}(\bar{\zeta}_i, \bar{\zeta}_j)\,z_i\,z_j\Big].$$

Since W is orthogonal to $\bar{W}$ with respect to $\hat{p}_{(x,\xi)}$, it follows that W is self-orthogonal with respect to $\tilde{p}_{(x,\xi)}$. Thus, the first term above vanishes, and we have

$$(2.140) \qquad \tilde{p}_{(x,\xi)}(\underline{s}, \underline{t}) = \frac{1}{2}\Big[\sum_{i,j} \tilde{p}_{(x,\xi)}(\zeta_i, \bar{\zeta}_j)\,\bar{z}_i\,z_j + \sum_{i,j} \tilde{p}_{(x,\xi)}(\bar{\zeta}_j, \zeta_i)\,z_j\,\bar{z}_i + \sum_{i,j} \tilde{p}_{(x,\xi)}(\bar{\zeta}_i, \bar{\zeta}_j)\,z_i\,z_j\Big].$$

Since $\tilde{p}_{(x,\xi)}$ is symmetric on $N(\Sigma)_{(x,\xi)}$ its extension to $N(\Sigma)_{(x,\xi)} \otimes \mathbb{C}$ is also symmetric, and so the expression (2.140) for $\tilde{p}_{(x,\xi)}$ is in "symmetric form". Thus, using (2.138) we get

$$(2.141) \qquad \widetilde{P}_{(x,\xi)} = \frac{1}{2}\Big[\sum_{i,j} \widetilde{p}_{(x,\xi)}(\zeta_i, \overline{\zeta}_j)\, Z_i^* Z_j + \sum_{i,j} \widetilde{p}_{(x,\xi)}(\zeta_i, \overline{\zeta}_j)\, Z_j Z_i^* - \sum_{i,j} \widetilde{p}_{(x,\xi)}(\overline{\zeta}_i, \overline{\zeta}_j)\, Z_i Z_j\Big] .$$

where

$$(2.142) \qquad Z_i \equiv D_{s_i} - \sqrt{-1}\, s_i \quad (\text{and so } Z_i^* = D_{s_i} + \sqrt{-1}\, s_i) .$$

Since $[Z_j, Z_i^*] = 2\delta_{ij}$ it follows that if we define N to be

$$(2.143) \qquad N \equiv \frac{1}{2}\Big[2 \sum_{i,j} \widetilde{p}_{(x,\xi)}(\zeta_i, \overline{\zeta}_j)\, Z_i^* Z_j - \sum_{i,j} \widetilde{p}_{(x,\xi)}(\overline{\zeta}_i, \overline{\zeta}_j)\, Z_i Z_j\Big]$$

then

$$(2.144) \qquad \widetilde{P}_{(x,\xi)} = N + \sum_i \widetilde{p}_{(x,\xi)}(\zeta_i, \overline{\zeta}_i) .$$

We next need a lemma. (Compare [4] section 4) .

<u>Lemma 2.45</u>. Let Θ be the $n \times n$ matrix whose entries are given by $\Theta_{ij} \equiv \widetilde{p}_{(x,\xi)}(\zeta_i, \overline{\zeta}_j)$. Then the eigenvalues

of Θ are precisely the eigenvalues of A corresponding to W (which, according to our assumption that $z = 1$, are precisely the eigenvalues of A with real part strictly positive).

Pf:

Using the basis $\zeta_1, \ldots, \zeta_n$ for W we see that if $u, v \in W$ have the expression $u = \Sigma u_i \zeta_i$ $v = \Sigma v_i \zeta_i$, $u_i, v_i \in \mathbb{C}$ then

$$(2.145) \quad \hat{p}_{(x,\xi)}(u, v) = \tilde{p}_{(x,\xi)}(u, \bar{v}) =$$

$$\sum_{i,j} \tilde{p}_{(x,\xi)}(\zeta_i, \bar{\zeta}_j)\, u_i \bar{v}_j = \sum_{i,j} \Theta_{ij}\, u_i \bar{v}_j .$$

But since $\zeta_1, \ldots, \zeta_n$ is an orthonormal basis of W with respect to $\hat{\omega}_{(x,\xi)}$ we have that

$$(2.146) \quad \hat{\omega}_{(x,\xi)}(u, v) = \sum_i u_i \bar{v}_i$$

Hence, if $(u_1, \ldots, u_n)$ is an eigenvalue of Θ^t, the transpose of Θ, with eigenvalue λ, it follows from (2.145) and (2.146) that the vector $u \in W$ satisfies

$$(2.147) \quad \hat{p}_{(x,\xi)}(u, v) = \lambda\, \hat{\omega}_{(x,\xi)}(u, v) \text{ for every } v \in W .$$

But since W is orthogonal to $\overline{W}$ both with respect to $\hat{p}_{(x,\xi)}$ and with respect to $\hat{\omega}_{(x,\xi)}$, and since W and $\overline{W}$ span $N(\Sigma)_{(x,\xi)} \otimes ¢$, it follows from (2.147) that

$$(2.148) \qquad \hat{p}_{(x,\xi)}(u, v) = \lambda\, \hat{\omega}_{(x,\xi)}(u, v) \quad \text{for every} \quad v \in N(\Sigma)_{(x,\xi)} \otimes ¢ \; .$$

Thus, by (2.106), it follows that λ is an eigenvalue of A (corresponding to an eigenvector in W). Conversely, reversing the above steps, it is clear that any eigenvalue of A corresponding to an eigenvector in W is an eigenvalue of Θ^t. A more careful analysis shows that multiplicities agree. Since the eigenvalues of the transpose of a matrix are the same as the eigenvalues of the matrix, the lemma follows.

QED

In view of Lemma 2.45 we have, in particular, that (2.144) may be written as

$$(2.149) \qquad \tilde{P}_{(x,\xi)} = N + \operatorname{trace}(\Theta) = N + \sum_{i=1}^{n} a_i$$

where the a_i are the eigenvalues of A corresponding to W.

We next set up the "concatenations". Using the fact that $[Z_i, Z_j] = 0$ and $[Z_i, Z_j^*] = 2\delta_{ij}$ we see that

$$(2.150) \qquad Z_k N = N Z_k + 2 \sum_j \tilde{p}_{(x,\xi)} (\zeta_k, \bar{\zeta}_j) Z_j$$

$$= N Z_k + 2 \sum_j \Theta_{kj} Z_j$$

From (2.143) we see that

$$(2.151) \qquad N^* = \tfrac{1}{2} [2 \sum_{i,j} \overline{\tilde{p}_{(x,\xi)} (\zeta_i, \bar{\zeta}_j)} Z_j^* Z_i$$

$$- \sum_{i,j} \overline{\tilde{p}_{(x,\xi)} (\bar{\zeta}_i, \bar{\zeta}_j)} Z_j^* Z_i^*] .$$

(Since we are only concerned with the action of our operators on $\mathcal{S}$, the Schwartz space, we do not have to worry about "boundary terms" when taking adjoints.)

Now using the fact that $[Z_i^*, Z_j^*] = 0$ and $[Z_i^*, Z_j] = -2\delta_{ij}$ we see that

$$(2.152) \qquad Z_k^* N^* = N^* Z_k^* - 2 \sum_j \overline{\tilde{p}_{(x,\xi)} (\zeta_k, \bar{\zeta}_j)} Z_j^*$$

$$= N^* Z_k^* - 2 \sum_j \overline{\Theta_{kj}} Z_j^*$$

Now letting I_n denote the $n \times n$ identity matrix,

$Z^{\#}$ the $n \times 1$ column matrix with $i1$ entry Z_i ,
$Z^{*\#}$ the $n \times 1$ column matrix with $i1$ entry Z_i^* ,
and $\overline{\Theta}$ the $n \times n$ matrix with ij entry $\overline{\Theta_{ij}}$ we can
rewrite (2.150) and (2.152) as

$$(2.153) \qquad Z^{\#} N = (NI_n + 2\,\Theta)\, Z^{\#}$$

and

$$(2.154) \qquad Z^{*\#} N^* = (N^* I_n - 2\,\overline{\Theta})\, Z^{*\#} \,.$$

We shall use (2.153) and (2.154) to determine the eigenvalues of N , and hence of $\widetilde{P}_{(x,\xi)}$. The procedure will be analoguous to that used earlier in the case $n = 1$. Indeed, (2.153) will play the role of (2.131), and (2.154) that of (2.130). The fact that (2.154) involves N^* rather than N will not cause us any difficulty, in view of the following lemma.

Lemma 2.46. Under the assumptions $n = 1$ and "conflicting influences" or $n > 1$, the eigenvalues of $\widetilde{P}^*_{(x,\xi)}$ are precisely the conjugates of the eigenvalues of $\widetilde{P}_{(x,\xi)}$, i.e., λ is an eigenvalue of $\widetilde{P}^*_{(x,\xi)}$ if and only if $\overline{\lambda}$ is an eigenvalue of $\widetilde{P}_{(x,\xi)}$.

Pf.

This follows from the fact (proved in §2.4) that the assumptions $n = 1$ and "conflicting influences" or $n > 1$ correspond precisely to the statement that $\mathrm{ind}(\tilde{P}_{(x,\xi)} - \lambda) = 0$ for any $\lambda \in \mathbb{C}$. More details may be found in the proof of Lemma 2.33.

QED

Since we are assuming conflicting influences, with $z = 1$, Lemma 2.42A and (2.149) imply

(2.155) The set $\{\mathrm{Re}\,\lambda \mid \lambda \text{ is an eigenvalue of } N\}$

is bounded below.

We need a further, easy, lemma.

Lemma 2.47. Let Q be a scalar differential operator, say from $\mathcal{S}$ to $\mathcal{S}$, and let M be an $n \times n$ matrix of complex numbers. Then there exists $\underline{f} = (f_1, \ldots, f_n) \in \mathcal{S} \oplus \cdots \oplus \mathcal{S}$ such that $\underline{f} \not\equiv 0$ and such that

(2.156) $(Q\, I_n - M)\, \underline{f} = 0$

if and only if there exists $g \in \mathcal{S}$ such that $g \not\equiv 0$ and such that

$$(2.157) \qquad (Q - \lambda)\, g = 0$$

for some eigenvalue λ of M.

<u>Pf</u>:

We may assume that M is in Jordan canonical form, for if C is any invertible $n \times n$ matrix of complex constants then

$$(2.158) \qquad C\, QI_n\, C^{-1} = QI_n$$

and so (2.156) is equivalent to

$$(2.159) \qquad (Q\, I_n - CMC^{-1})\, C\underline{f} = 0$$

and, since C is invertible, $C\underline{f} = 0$ if and only if $\underline{f} = 0$.

So, we can, without loss of generality write

$$(2.160) \qquad M = \begin{pmatrix} M_1 & & 0 \\ & \ddots & \\ 0 & & M_\ell \end{pmatrix}$$

each M_j being a Jordan block corresponding to an eigenvalue λ_j of M. (Of course there may be more than one block with a given eigenvalue.)

Write $\underline{f} = (\underline{f}_1,\ldots,\underline{f}_\ell)$, where the notation is self-explanatory. Of course $\underline{f} = 0$ if and only if, for every j , $\underline{f}_j = 0$, and $(Q - M)\underline{f} = 0$ if any only if, for every j , $(Q - M_j) f_j = 0$. Thus, there exists $\underline{f} \neq 0$ such that (2.156) holds if and only if for some j there exists $\underline{f}_j \neq 0$ such that $(Q - M_j) \underline{f}_j = 0$.

Let λ_j be the eigenvalue corresponding to M_j , so that

$$(2.161) \qquad M_j = \begin{pmatrix} \lambda_j & 1 & & \bigcirc \\ & \lambda_j & 1 & \\ & & & 1 \\ \bigcirc & & & \lambda_j \end{pmatrix} ,$$

Let $\underline{f}_j = (f^1,\ldots,f^r)$. Then, of course, $\underline{f}_j = 0$ if and only if, for every ℓ , $f^\ell \neq 0$. Assuming $\underline{f}_j \neq 0$, let ℓ_0 be the largest ℓ such that $f^\ell \neq 0$. Then it is clear from (2.161) that if $(Q - M_j)(\underline{f}_j) = 0$ then $(Q - \lambda_j)f^{\ell_0} = 0$. Conversely, if $g \neq 0$ satisfies $(Q - \lambda_j)\, g = 0$, then

$$(Q - M_j) \begin{pmatrix} 0 \\ \vdots \\ 0 \\ g \end{pmatrix} = 0 .$$

QED

Recall from Lemma 2.45 that the eigenvalues of Θ are precisely the eigenvalues of $a_1,\ldots,a_n$ of A with strictly positive real part. Now suppose that v_0 is an eigenvector of N with eigenvalue λ , i.e., $v_o \neq 0$ and $Nv_o = \lambda v_o$. Then, applying (2.153) we get that $\lambda Z^{\#} v_o = (NI_n + 2 \Theta) Z^{\#} v_o$. That is,

$$(2.162) \qquad [NI_n - (\lambda - 2\Theta)] Z^{\#} v_o = 0 .$$

Thus, by Lemma 2.47, if $Z^{\#} v_o \neq 0$, i.e., if $Z_j v_o \neq 0$ for some $j = 1,\ldots,n$, then there exists $v_1 \neq 0$ such that

$$(2.163) \qquad [N - (\lambda - 2a_i)] v_1 = 0$$

for some $i = 1,\ldots,n$. That is, v_1 is an eigenvector of N with eigenvalue $\lambda - 2a_i$.

Iterating this process we see that, having constructed eigenfunctions $v_o,\ldots,v_k$ of N with eigenvalues $\lambda - 2\sum_{i=1}^{n} r_{ik} a_i$, r_{ik} being non-negative integers such that $\sum_{i=1}^{n} r_{ik} = k$, we can, unless $Z^{\#} v_k = 0$, i.e., <u>unless $Z_j v_k = 0$ for every $j = 1,\ldots,n$</u>, construct an eigenfunction v_{k+1} of N with eigenvalue $\lambda - 2\sum_{i=1}^{n} r_{i(k+1)} a_i$, where $r_{i(k+1)}$ are non-negative integers such that $\sum_{i=1}^{n} r_{i(k+1)} = k+1$.

Since the a_i all have strictly positive real part it follows that the sequence of eigenfunctions $v_o,\ldots v_k,\ldots$

must terminate at some position k, for otherwise the boundedness below property (2.155) would be violated. That is, we have $v_k \neq 0$ satisfying

$$(2.164) \qquad Z_j v_k = 0 \quad \text{for every } j = 1,\dots,n$$

and

$$(2.165) \qquad Nv_k = (\lambda - 2 \sum_{i=1}^{n} r_{ik} a_i) v_k \quad .$$

But, in view of the definition (2.143) of N, it follows from (2.164) that $Nv_k = 0$. Since $v_k \neq 0$ it follows from (2.165) that $\lambda - 2 \sum_{i=1}^{n} r_{ik} a_i = 0$. Thus we have shown

(2.166) Every eigenvalue of N is of the form

$$2 \sum_{i=1}^{n} a_i n_i$$

where the n_i are non-negative integers.

We want to prove the converse of (2.166), namely that for any non-negative integers n_i, $2 \sum_{i=1}^{n} a_i n_i$ is an eigenvalue of N. To do this we must first discuss the Hermite functions.

Let s denote the variable for $L^2(\mathbb{R})$ and define Z and Z^* by $Z = D_s - \sqrt{-1}\, s$ and $Z^* = D_s + \sqrt{-1}\, s$.

Then it is well-known that there is a complete orthonormal basis of eigenfunctions, $H_0(s)$, $H_1(s)$, $H_2(s)$,... (each lying in $\mathscr{S}$) of the operator Z^*Z . Indeed, the following relations are satisfied:

$$(2.167)\qquad Z(H_k) = \begin{cases} \frac{1}{\sqrt{-1}}\sqrt{2k}\ H_{k-1} & \text{if } k \text{ is a positive integer} \\ 0 & \text{if } k = 0 \end{cases}$$

$$(2.168)\qquad Z^*(H_k) = -\frac{1}{\sqrt{-1}}\sqrt{2(k+1)}\ H_{k+1} \quad \text{for every}$$

non-negative integer k .

From (2.167) and (2.168) it follows that

$$(2.169)\qquad Z^*Z\ (H_k) = 2k\ H_k \quad \text{for every non-negative integer } k \ .$$

Of course, the Hermite functions H_k are, up to constant multiples, precisely the eigenfunctions obtained via the method of concatenations. Indeed, if we begin with $H_0(s)$, which equals $\pi^{-1/4}\ e^{-s^2/2}$, then the succeeding H_k may be obtained recursively from (2.168).

Taking products of Hermite functions we obtain an orthonormal basis for $L^2(s_1,\dots,s_n)$. More precisely, for any n-tuple $\underline{k} = (k_1,\dots,k_n)$ of non-negative integers we define $H_{\underline{k}}$ by

(2.170) $$H_{\underline{k}}(s_1,\ldots,s_n) = H_{k_1}(s_1)\, H_{k_2}(s_2) \cdots H_{k_n}(s_n),$$

and the $H_{\underline{k}}$ form an orthonormal basis for $L^2(s_1,\ldots,s_n)$.

Then from (2.167) and (2.168) we see that

(2.171) $$Z_j(H_{\underline{k}}) = \begin{cases} \frac{1}{\sqrt{-1}} \sqrt{2k_j}\; H_{\underline{k}-1_j} & \text{if } k_j \text{ is positive} \\ 0 & \text{if } k_j = 0 \end{cases}$$

and

(2.172) $$Z_j^*(H_{\underline{k}}) = -\frac{1}{\sqrt{-1}} \sqrt{2(k_j+1)}\; H_{\underline{k}+1_j} \quad .$$

Here 1_j denotes the n-tuple having 1 in the j-th position and 0 in all the other positions.

By the <u>degree</u> of $H_{\underline{k}}$ we mean $\sum_{i=1}^{n} k_i$. Notice, for example, that Z_j lowers degree by 1 (if $k_j \neq 0$) and that Z_j^* raises degree by 1 .

We shall prove the converse of (2.166) by using (2.154) to show

(2.173) For any non-negative integers n_i, $2\sum_{i=1}^{n} \bar{a}_i n_i$ is an eigenvalue of N^* .

Indeed, applying Lemma 2.46 we can conclude from (2.173) that

(2.174) For any non-negative n_i, $2\sum_{i=1}^{n} a_i n_i$ is an eigenvalue of N.

(Lemma 2.46 is certainly applicable here, for $\tilde{P}_{(x,\xi)}$ differs from N only by a constant, $\sum_{i=1}^{n} a_i$.)

First we prove (2.173) for $n_i = 0$, $i = 1,\ldots,n$, i.e., we prove that 0 is an eigenvalue of N^*. In fact, since, by (2.171), $Z_j(H_{\underline{o}}) = 0$ for every j it follows from the definition (2.143) of N that $N(H_{\underline{o}}) = 0$. In particular, 0 is an eigenvalue of N, and hence, by Lemma 2.46, an eigenvalue of N^*.

Since we know (2.173) holds for $n_i = 0$, $i = 1,\ldots,n$, the proof of (2.173) will be complete if we prove the inductive step:

(2.175) If λ is an eigenvalue of N^* then so is $\lambda + 2\bar{a}_i$ for every $i = 1,\ldots,n$.

Let v be an eigenfunction corresponding to the eigenvalue λ. Since the $\bar{a}_i$ are precisely the eigenvalues of $\bar{\Theta}$, an examination of the proof of Lemma 2.47 reveals that (2.175) follows from the commutation relation (2.154)

provided we can show that the component of $Z^{*\#}v$ corresponding to each generalized eigenspace of $\overline{\Theta}$ is non-zero. If C is an invertible $n \times n$ matrix such that $C\,\overline{\Theta}\,C^{-1}$ is in Jordan canonical form then $Z^{*\#}v$ will certainly have a non-zero component corresponding to each generalized eigenspace of $\overline{\Theta}$ if each entry of the $n \times 1$ column matrix $C\,Z^{*\#}v$ is non-zero , i.e. , if $\sum_{j=1}^{n} c_{ij}\, Z_j^{*} v$ is non-zero for every $i = 1 , \ldots , n$. Since C is invertible none of its rows is identically 0 . Thus, once we prove the following lemma the inductive step (2.175) , and hence (2.173) and (2.174) will follow.

<u>Lemma 2.48.</u> Let $(c_1, \ldots, c_n)$ be an n-tuple of complex numbers not identically zero. Then the operator $\sum_{j=1}^{n} c_j Z_j^{*} : \mathcal{S} \longrightarrow \mathcal{S}$ is injective.

<u>Remark:</u> The use of $\mathcal{S}$ is not essential; indeed any subspace of L^2 which $\sum_{j=1}^{n} c_j Z_j^{*}$ maps into L^2 will do. However, since we know that all the eigenvectors of N lie in $\mathcal{S}$, Lemma 2.48 is sufficiently general for our purposes.

<u>Pf:</u>

Let f be a <u>non-zero</u> element of $\mathcal{S}$. Since the Hermite functions form an orthonormal basis for L^2 , we can write $f = \sum_{|\underline{k}| \geq 0}^{\infty} a_{\underline{k}} H_{\underline{k}}$, where not all the $a_{\underline{k}}$ equal 0. Here $|\underline{k}| = \sum_{i=1}^{n} k_i$, the degree of $H_{\underline{k}}$, (Remember that $k_i \geq 0$ for every i.) Let ℓ be the lowest degree for which terms with non-zero coefficients occur. Writing $f : \sum_{|\underline{k}|=\ell} a_{\underline{k}} H_{\underline{k}} + \sum_{|\underline{k}|>\ell}^{\infty} a_{\underline{k}} H_{\underline{k}}$

we shall show that $\sum_{j=1}^{n} c_j Z_j^*(\sum_{|\underline{k}|=\ell} a_{\underline{k}} H_{\underline{k}})$ is orthogonal, with respect to the L^2 inner product $\langle\ ,\ \rangle$ to $\sum_{j=1}^{n} c_j Z_j^* (\sum_{|\underline{k}|>\ell} a_{\underline{k}} H_{\underline{k}})$. Hence to prove Lemma 2.48 it will be sufficient to show that

$$\sum_{j=1}^{n} c_j Z_j^* (\sum_{|\underline{k}|=\ell} a_{\underline{k}} H_{\underline{k}}) \neq 0 . \tag{2.176}$$

To prove the above orthogonality assumption it suffices, since $\sum_{|\underline{k}|=\ell} a_{\underline{k}} H_{\underline{k}}$ is a finite sum, to show that

$$< Z_i^* (H_{\underline{\ell}}) , \ Z_j^* (\sum_{|\underline{k}|>\ell}^{\infty} a_{\underline{k}} H_{\underline{k}}) > = 0 \tag{2.177}$$

for every i,j and for every $\underline{\ell}$ such that $|\underline{\ell}| = \ell$.

But $< Z_i^* (H_{\underline{\ell}}) , Z_j^* (\sum_{|\underline{k}|>\ell}^{\infty} a_{\underline{k}} H_{\underline{k}}) > = < Z_j Z_i^*(H_{\underline{\ell}}) , \sum_{|\underline{k}|>\ell}^{\infty} a_{\underline{k}} H_{\underline{k}} >$. By (2.171) - (2.172), $Z_j Z_i^*(H_{\underline{\ell}})$ has degree ℓ . Hence, by the orthogonality of the Hermite functions, the above inner product equals 0 , i.e. , (2.177) holds.

We next prove (2.176) . (Remember, at least one $a_{\underline{k}}$ with $|\underline{k}| = \ell$ does not equal 0 .) Let j_o be the largest j such that $c_j \neq 0$. Let I_1 be the set of those $\underline{k}$ with $|\underline{k}| = \ell$ and $a_{\underline{k}} \neq 0$ having minimal k_1 . Having defined I_r , define I_{r+1} to be the set of those $\underline{k}$ in I_r having minimal k_{r+1} . Let $\underline{k}^o$ be any element of I_{j_o} . Then I claim that $c_{j_o} Z_{j_o}^* (H_{\underline{k}^o})$ (which $\neq 0$) is orthogonal to all other terms of the form $c_j Z_j^*(H_{\underline{k}})$ with $|\underline{k}| = \ell$ and $a_{\underline{k}} \neq 0$. If $j = j_o$ it is clear from (2.172) that, unless $\underline{k} = \underline{k}^o$, $c_{j_o} Z_{j_o}^* (H_{\underline{k}^o}) \perp c_j Z_j^*(H_{\underline{k}})$. Since j_o is the largest j such that $c_j \neq 0$, it suffices to consider $j < j_o$. Now if $c_{j_o} Z_{j_o}^* (H_{\underline{k}^o})$ is not orthogonal to $c_j Z_j^*(H_{\underline{k}})$ it follows that, for $r < j$, $\underline{k}$ must lie in I_r . Indeed this follows from the fact that $\underline{k}^o \in I_r$ and from the fact that neither Z_j^* nor $Z_{j_o}^*$ affect any part of the indices $\underline{k}$, $\underline{k}^o$, respectively, prior to the j-th place. But since $\underline{k}$ and $\underline{k}^o$ both lie in I_r for $r < j$, and since $\underline{k}^o$ lies in I_j it follows from the definition of I_j that $k_j^o \leqq k_j$. But Z_j^* raises k_j by one , and since $j < j_o$, $Z_{j_o}^*$ does not affect k_j^o . Thus, the indices of $c_{j_o} Z_{j_o}^* (H_{\underline{k}^o})$ and $c_j Z_j^*(H_{\underline{k}})$ differ in the j-th place, and hence these two terms are orthogonal. Since $c_{j_o} Z^*_{j_o}(H_{\underline{k}^o})$, which $\neq 0$, is orthogonal to all other terms of the form $c_j Z^*_j(H_{\underline{k}})$ with $|\underline{k}| = \ell$ and $a_{\underline{k}} \neq 0$, it follows that (2.176), and hence Lemma 2.48. holds.

QED

Thus, as stated earlier, (2.174) holds. This together with (2.166) and (2.149) proves Prop. 2.44.

QED

Remark: Since our test-operator $\tilde{P}_{(x,\xi)}$ has "constant-coefficients" we do not need to introduce the additional "ellipticity" assumption of ([4], section 4) in order to set up the "concatenations" (2.153) and (2.154).

We shall conclude this section with a brief discussion of the eigenvalues of $\tilde{P}_{(x,\xi)}$ (for p real) from the viewpoint of Maslov asymptotics ([19], [20], [5]). We shall see that the eigenvalues of $\tilde{P}_{(x,\xi)}$ are picked out by appropriate Lagrangian submanifolds of the symplectic manifold $N(\Sigma)_{(x,\xi)}$. (Recall that an n-dimensional submanifold Λ of a $2n$-dimensional symplectic manifold (M, ω) is Lagrangian if the tangent space to Λ at any point of Λ is self-annihilating with respect to ω.)

As we saw in §2.1 our asymptotic structure is somewhat different from that of Maslov. Whereas Maslov deals with a fixed symplectic manifold, T^*Y for some fixed Y, and studies asymptotic behavior in $\frac{1}{h}$ as "Plank's constant" h goes to 0, we deal with a family $N(\Sigma) \longrightarrow \Sigma$, consisting of a fibering, each fiber being a symplectic manifold, over a space Σ with an $\mathbb{R}^+$ action (in this case given by $\langle \rho, (x,\xi)\rangle \longmapsto (x, \rho\xi)$, for $\rho \in \mathbb{R}^+$ and $(x,\xi) \in \Sigma)$, and we study asymptotic behavior in ρ as ρ goes to $+\infty$.

Of course, from the view-point of asymptotics our test-operators are rather special. Indeed, as we pointed out in §2.1, if p is homogeneous of degree m, then our test-operators $\tilde{P}_{(x,\xi)} + \sigma_{sub}(P)|_{(x,\xi)}$ are homogeneous of degree $m-1$ with respect to the $\mathbb{R}^+$ action on Σ, in the sense, for example, that the eigenvalues $\lambda_j|_{(x,\rho\xi)}$ of $\tilde{P}_{(x,\rho\xi)} + \sigma_{sub}(P)|_{(x,\rho\xi)}$ are $\rho^{m-1}\lambda_j|_{(x,\xi)}$. Thus, there are no lower-order terms in the asymptotic expansion $\rho^{m-1}\left(\sum_{i=0}^{\infty} \lambda_{ji}\, \rho^{-i}\right)$. Hence, the Maslov theory, which deals in general with the leading term of the asymptotic expansion, in our case yields the exact eigenvalues. Moreover, the Maslov theory does not, in general, provide approximations to <u>all</u> the eigenvalues of the operator in question. However, again because of the special nature of $\tilde{P}_{(x,\xi)}$, ($\tilde{P}_{(x,\xi)}$ being simply a sum of harmonic oscillators) all the eigenvalues of $\tilde{P}_{(x,\xi)}$ are obtained.

Let $E\,\rho^{m-1}$ be an eigenvalue of $\tilde{P}_{(x,\rho\xi)}$. Then we will see that for an appropriate compact Lagrangian submanifold $\Lambda_{(x,\rho\xi)}$ of $N(\Sigma)_{(x,\rho\xi)}$ contained in $\{v \in N(\Sigma)_{(x,\rho\xi)} | \tilde{p}_{(x,\rho\xi)}(v, v) = E\,\rho^{m-1}\}$ the following quantization conditions of Maslov holds:

$$(2.178) \qquad \frac{1}{2\pi}\int_{\gamma} \eta_{(x,\rho\xi)} = \frac{1}{4}\,\text{ind}\,\gamma \mod 1$$

$$\text{for every } \gamma \in H_1(\Lambda_{(x,\rho\xi)}, \mathbb{Z}) .$$

Here $\eta_{(x,\rho\xi)}$ is a 1-form on $N(\Sigma)_{(x,\rho\xi)}$ such that $d\eta_{(x,\rho\xi)} = \omega_{(x,\rho\xi)}$. (Since $N(\Sigma)_{(x,\rho\xi)} \cong \mathbb{R}^{2n}$, closed is the same as exact, and so if $d\eta_1 = d\eta_2$, $\eta_1 - \eta_2 = d\varphi$ for some function φ. Thus, there is no ambiguity in (2.178) stemming from the choice of $\eta_{(x,\rho\xi)}$.) Ind γ denotes the Maslov index of γ. This is defined in terms of a fixed "polarization" for $N(\Sigma)_{(x,\rho\xi)}$, i.e., a fixed separation into s-variables and t-variables for some choice of symplectic linear coordinates $s_1,\ldots,s_n$, $t_1,\ldots,t_n$ for $N(\Sigma)_{(x,\rho\xi)}$. Indeed, Ind γ is the intersection number of γ with the singular cycle of $\Lambda_{(x,\rho\xi)}$ with respect to the projection $\pi : (\underline{s}, \underline{t}) \longmapsto \underline{s}$, i.e., the set of points of $\Lambda_{(x,\rho\xi)}$ at which $\pi \,|\, \Lambda_{(x,\rho\xi)}$, the restriction of π to $\Lambda_{(x,\rho\xi)}$, fails to have maximal rank. I do not know (except in the case $n = 1$, where this is clear) whether, for the Lagrangian manifolds considered below, ind γ is independent of the choice of polarization. Thus when we verify (2.178) we shall use a fixed polarization in discussing ind γ.

We begin with the case $n = 1$. Fix $(x,\xi) \in \Sigma$ and choose local parameters u, v, homogeneous of degree $\frac{1}{2}$, for Σ near (x,ξ), satisfying the canonical commutation relations at (x,ξ), and hence, because of the homogeneity, also at $(x,\rho\xi)$ for every $\rho > 0$. Thus, using du, dv as local frames for $N(\Sigma)$, with corresponding symplectic coordinates s,t, we can write $\tilde{p}_{(x,\rho\xi)}(s, t) \equiv \tilde{p}_{(x,\rho\xi)}(sdu + tdv, sdu + tdv)$ as

$$(2.179) \qquad \tilde{p}_{(x,\rho\xi)}(s,t) = a_{(x,\rho\xi)}\, t^2 + 2c_{(x,\rho\xi)}\, st + b_{(x,\rho\xi)}\, s^2 .$$

Write a, b, c for $a_{(x,\xi)}, b_{(x,\xi)}, c_{(x,\xi)}$, respectively. Because of the homogeneity of p, u, v we have (see (2.40) and (2.42)) that

$$(2.180) \qquad a_{(x,\ \xi)} = a\,\rho^{m-1}, \quad b_{(x,\rho\xi)}\ b\,\rho^{m-1}, \quad c_{(x,\rho\xi)} = c\,\rho^{m-1} .$$

Since $\tilde{p}_{(x,\rho\xi)}$ is definite (assume positive definite), the set $\{(s, t) \mid \tilde{p}_{(x,\rho\xi)}(s, t) = E\,\rho^{m-1}\}$ (which by (2.180) equals $\{(s, t) \mid \tilde{p}_{(x,\xi)}(s, t) = E\}$) is a 1-dimensional compact manifold in (s, t) space, in fact an ellipse. This manifold is Lagrangian since it has proper dimension and since its tangent space at any point, being 1-dimensional, is self-annihilating with respect to the anti-symmetric form ω. (Of course it is the only compact Lagrangian manifold contained in the ellipse). We take this manifold to be $\Lambda_{(x,\rho\xi)}$ and check whether (2.178) holds. It is certainly enough to test a basis of homology classes in $H_1(\Lambda_{(x,\rho\xi)}, \mathbb{Z})$, i.e., in this case it suffices to test the curve γ which goes around the ellipse once clockwise.

It is easily verified that $\operatorname{ind} \gamma = 2$, for the singular cycle of $\Lambda_{(x,\rho\xi)}$ consists precisely of the two points having vertical tangents, i.e., tangents parallel to the t-axis, and at each of these points $\frac{ds}{dt}$ changes sign from $-$ to $+$ as we run along γ. Also, since (see (2.39)) $\omega_{(x,\rho\xi)} = dt \wedge ds$, we can take $\eta_{(x,\rho\xi)}$ to be $t\,ds$. So, (2.178) becomes

$$(2.181) \qquad \frac{1}{2\pi} \int_\gamma t\, ds = \frac{1}{2} \mod 1$$

Applying Stokes' theorem, we can write this as

$$(2.182) \qquad \frac{1}{2\pi} (\text{Area of ellipse}) = \frac{1}{2} \mod 1$$

But the ellipse $\{(s, t) \mid \tilde{p}_{(x,\xi)}(s, t) = E\}$ is the image under T^{-1} of the circle of radius $E^{\frac{1}{2}}$ in (s, t) space, where T is the square root of the positive definite matrix $\begin{pmatrix} b & c \\ c & a \end{pmatrix}$. Thus

(2.183) $$\text{Area of ellipse} = \pi E/\sqrt{ab - c^2}$$

and so (2.182) becomes

(2.184) $$E = \sqrt{ab - c^2} \cdot (2n + 1)$$ for some (necessarily nonnegative) integer.

Of course, if $\tilde{p}_{(x,\xi)}$ is expressed in "diagonal form" , with $a = b$, $c = 0$, then (because of the positive definiteness) $\sqrt{ab - c^2} = a$, and thus the eigenvalues E of $\tilde{P}_{(x,\xi)}$ obtained via (2.184) are the same as those obtained earlier. It may be shown directly that $\sqrt{ab - c^2}$ is an invariant associated to $\tilde{p}_{(x,\xi)}$ and $\omega_{(x,\xi)}$. In fact if, in analogy with (2.106), we define the linear map $B : N(\Sigma)_{(x,\xi)} \longrightarrow N(\Sigma)_{(x,\xi)}$ (notice we do not complexify) by

(2.185) $$\omega_{(x,\xi)}(Bu, v) = \tilde{p}_{(x,\xi)}(u, v)$$

for every $u, v \in N(\Sigma)_{(x,\xi)}$

then it is easy to verify that $ab - c^2 = \det B$.

To treat the case $n > 1$ it is convenient to choose symplectic coordinates $s_1, \ldots, s_n$, $t_1, \ldots, t_n$ for which $\tilde{p}_{(x,\rho\xi)}$ is in diagonal form:

$$(2.186) \qquad \tilde{p}_{(x,\rho\xi)} (\underline{s}, \underline{t}) = \sum_{i=1}^{n} a_i|_{(x,\rho\xi)} (s_i^2 + t_i^2) \ .$$

Of course, if we write a_i for $a_i|_{(x,\xi)}$ then, choosing frames as in the case $n = 1$, we have the homogeneity property (2.180), and as above, the set $\{(\underline{s}, \underline{t}) | \tilde{p}_{(x,\rho\xi)}(\underline{s}, \underline{t}) = E\rho^{m-1}\}$ equals the set $\{(\underline{s}, \underline{t}) | \tilde{p}_{(x,\xi)} (\underline{s}, \underline{t}) = E\}$. The compact Lagrangian manifolds contained in $\{(\underline{s}, \underline{t}) | \tilde{p}_{(x,\xi)} (\underline{s}, \underline{t}) = E\}$ which we shall consider are the tori.

$$(2.187) \qquad \Lambda_{(x,\xi)} (E_1, \ldots, E_n) = \{(\underline{s}, \underline{t}) \mid a_i (s_i^2 + t_i^2) = E_i, \ i = 1, \ldots, n\}$$

where the E_i are non-negative numbers such that $\sum_{i=1}^{n} E_i = E$. We shall determine which $\Lambda_{(x,\xi)}(E_1, \ldots, E_n)$ satisfy the quantization conditions (2.178). As a basis for $H_1 (\Lambda_{(x,\xi)}, \mathbb{Z})$ we may take the curves γ_i, $i = 1, \ldots, n$, where γ_i winds once "clock-wise" about the circle $a_i (s_i^2 + t_i^2) = E_i$, s_j, t_j being constant for $j \neq i$. As in the case $n = 1$, it is easy to verify that $\operatorname{ind} \gamma_i = 2$ for every i. Since $\eta_{(x,\rho\xi)} = \sum_{i=1}^{n} t_i \, ds_i$ it follows that (2.178) takes the form (where we view

γ_i as a curve in (s_i, t_i) space)

$$(2.188) \qquad \frac{1}{2\pi} \int_{\gamma_i} t_i ds_i = 1/2 \bmod 1 \quad \text{for every} \quad i = 1,\ldots,n.$$

But this is simply n copies of (2.181), and so is equivalent to n copies of (2.184), i.e., to

$$(2.189) \qquad E_i = a_i (2n_i + 1) \quad \text{for some non-negative integer } n_i.$$

Since $E = \sum_{i=1}^{n} E_i$, this means of course that

$$(2.190) \quad E = \sum_{i=1}^{n} a_i (2n_i + 1) \text{ for non-negative integers } n_i .$$

As we saw earlier, these are precisely all the eigenvalues of $\tilde{\tilde{P}}_{(x,\xi)}$. We note that Maslov actually associates to each of the Lagrangian manifolds $\Lambda_{(x,\xi)}(a_1(2n_1 + 1),\ldots, a_n(2n_n + 1))$ an approximate eigenfunction; however, we shall not discuss this.

§3. Example: Poincaré complexes $(\bar{\partial}_b)$.

In this section we use the results of §2. to obtain a criterion for hypoellipticity with loss of one derivative for the Laplacians associated to a Poincaré complex of first-order pseudo-differential operators. We assume that the Levi-form is non-degenerate, which insures that the hypotheses of Theorem 2.4 hold. The principal symbol of each of the Laplacians Δ_k is real, so we know from §2,5 that, for each $(x,\xi) \in \Sigma$, $(\tilde{\Delta}_k)_{(x,\xi)}$ is a sum of independent harmonic oscillators. We shall see that the "weights" are precisely $\frac{|\lambda_i|}{2}$, $i = 1,\ldots,q$, where the λ_i are the eigenvalues of the Levi-form. Notice that $(\tilde{\Delta}_k)_{(x,\xi)}$ does not depend on k, for the principal symbol of Δ_k does not vary with k. The "lower-order" parts, $\sigma_{sub}(\Delta_k)_{(x,\xi)}$, of our test-operator also are expressed in terms of the λ_i, and do depend on k. We shall see that for these test-operators the eigenvalue criterion of Theorem 2.4 translates into the Levi-form criterion for $\frac{1}{2}$-subellipticity of the original Poincaré complex. In particular, we get the well-known Levi-form criterion for $\frac{1}{2}$-subellipticity of the tangential Cauchy-Riemann complex, $\bar{\partial}_b$. (For a discussion of the $\bar{\partial}_b$ complex, see [8].) The $\frac{1}{2}$-subellipticity criterion for Poincaré complexes may be found, for example, in [15], [22], or [23]. [22] also contains a discussion relating the Levi-form treated here to the usual one.) Notice that the "discreteness" of the hypoellipticity criterion is "disguised" since the lower-order symbol of Δ_k bears such a special relationship to the principal symbol.

It is quite likely that, by combining the micro-localizability results of Boutet de Monvel and Treves [4] for hypoellipticity with loss of one derivative with the canonical microlocal decomposition ([15], [22], [23]) of a complex of first-order differential operators into a direct sum of an exact complex and a Poincaré complex, it should be possible to get a Levi-form criterion for hypo-ellipticity with loss of one derivative for the Laplacians associated to any first-order Spencer complex with simple characteristics and non-degenerate Levi-form. In the process one would have to show that, just as $\frac{1}{2}$-subelliptic-ticity of the complex does not depend on the choice of Hermitian metrics for the bundles, neither does hypo-ellipticity with loss of one derivative for the associated Laplacians; for the canonical microlocal decomposition of the complex does not bear any relation to the Hermitian metrics of the bundles. In particular, the decomposition need not be orthogonal. We let the matter rest here, since we have not attempted to carry out a proof of the above. Before proceeding to the details of our computation, we point out that in the case of $\overline{\partial}_b$ the test-operators occur in a somewhat different manner (as the operators $\pi_\lambda(L_\alpha)$) in the work ([9], [10]) of Folland and Stein.

By taking an appropriate choice of Hermitian metrics, they arrange that the "weights" of the harmonic osciallators which appear are all equal. A computation for $\bar{\partial}_b$ similar to that presented below (with metrics chosen so that the analogues of our "weights" are all ± 1) has been done by Boutet de Monvel and Treves. This is indicated in [3] . See also [2] .

We begin by recalling a definition ([15], [22], [23]) .

Definition 3.1. Let E^o be a vector bundle over a manifold M , and suppose that $P_1,\ldots,P_q$ are commuting first-order pseudo-differential operators from E^o to E^o . Then the associated Poincaré complex $\{P, E^i\}$ is defined as follows: Let W be a q-dimensional vector space over $\mathbb{C}$ with distinguished basis $w_1,\ldots,w_q$. Then for $i = 0, \ldots,q$ let

$$(3.1) \qquad E^i = E^o \otimes \Lambda^i W$$

and define $P : E^i \longrightarrow E^{i+1}$ by

$$(3.2) \qquad P\,(\sum_I f_I \otimes w_I) = \sum_{j,I} P_j f_i \otimes (w_j \wedge w_I)$$

Remark. Since $P_i\, P_j = P_j\, P_i$ and $w_i \wedge w_j = -w_j \wedge w_i$ it follows that $\{P, E^i\}$ is a complex, i.e., that $P^2 = 0$.

Next, assume that Hermitian metrics have been chosen for the bundles E^i (and that a Riemannian metric has been chosen for M), so that the formal adjoint $P^* : E^i \longrightarrow E^{i-1}$ is defined.

Definition 3.2. The Laplacians $\Delta_i : E^i \longrightarrow E^i$, $i = 0,\ldots,q$, associated to $\{P, E^i\}$ are the second-order pseudo-differential operators defined by

$$(3.3) \qquad \Delta_i = PP^* + P^*P \quad ,$$

where the first P^* denotes the adjoint of $P : E^{i-1} \longrightarrow E^i$, and the second P^* denotes the adjoint of $P : E^i \longrightarrow E^{i+1}$.

We shall assume that $\{P, E^i\}$ has "simple characteristics", i.e., the principal symbols p_i of the commuting operators P_i are scalar maps, and, in fact, the $\{p_i\}$ are a set of parameters for the complex characteristic variety $\mathcal{U}$ in a neighborhood of the real characteristic point (x_o, ξ_o) . That is, we assume

(3.4) $p_i(x, \xi)$ is holomorphic in ζ , homogeneous of degree 1 in ζ , and smooth in x .

(3.5) $\text{grad}_\zeta\, p_i(x_o, \xi_o)$, $i = 1,\ldots,q$, are linearly independent (over $\mathbb{C}$) .

Of course, in a neighborhood of (x_o, ξ_o) , $\mathcal{U}$ is equal to the set of points where $p_1 = \cdots = p_q = 0$. If we let Σ equal $\mathcal{U} \cap T^*_{\mathbb{R}}$ then it is easy to check that Σ is , as well as the real characteristic variety of $\{P, E^i\}$, also the real characteristic variety of Δ_i , for every i .

At a point $(x_o, \xi_o) \in \Sigma$ the Levi-form is defined to be the $q \times q$ Hermitian matrix $\frac{1}{\sqrt{-1}} \{p_i, \overline{p}_j\}|_{(x_o, \xi_o)}$. It is known that, under the assumptions (3.4), (3.5), the signature of the Levi-form is independent of the choice of parameters for $\mathcal{U}$. Indeed, the Levi-form is, essentially, the restriction to the n-dimensional "holomorphic" supspace of $N(\Sigma)_{(x_o, \xi_o)} \otimes \mathbb{C}$ of the form $\hat{\omega}_{(x_o, \xi_o)}$ of (2.101) . Strictly speaking, to talk of $N(\Sigma)$ we must first show that Σ is a manifold. We shall see below that this holds, in particular, if the Levi-form is non-degenerate. In fact, under this assumption, we will be able to conclude that if u_i, v_i are, respectively, the real and imaginary parts of p_i, then $du_1, \ldots, du_q$, $dv_1, \ldots, dv_q$ are linearly independent at $(x_o, \xi_o) \in \Sigma$, and thus form a set of parameters defining Σ , and, that, moreover, the form $\omega_{(x_o, \xi_o)}$ on $N(\Sigma)_{(x_o, \xi_o)}$ is non-degenerate.

Indeed, by assuming the Hermitian matrix

$\frac{1}{\sqrt{-1}} \{p_i, \overline{p}_j\}|_{(x_o, \xi_o)}$ is non-degenerate, we have, after diagonalizing by a unitary transformation, that

$$\frac{1}{\sqrt{-1}} \{p_i, \overline{p}_j\}|_{(x_o, \xi_o)} = \lambda_i \delta_{ij} , \tag{3.6}$$

where $\lambda_i > 0, i = 1, \ldots, r$; $\lambda_i < 0$, $i = r+1, \ldots, q$.

(That is, none of the λ_i equals 0.) But, since the P_i's commute, the principal symbol of the commutator vanishes, in particular on Σ. Thus we have

$$(3.7) \qquad \{p_i, p_j\}|_{(x_o,\xi_o)} = 0 .$$

Together, (3.6) and (3.7) are equivalent to

$$(3.8) \qquad \{u_i, u_j\}|_{(x_o,\xi_o)} = 0 \ , \quad \{v_i, v_j\}|_{(x_o,\xi_o)} = 0 \ ,$$

$$\{u_i, v_j\}|_{(x_o,\xi_o)} = \frac{-\lambda_i}{2} \delta_{ij} .$$

The linear independence at (x_o, ξ_o) of $du_1,\ldots,du_q$, $dv_1,\ldots,dv_q$ follows from (3.8) and the fact that none of the λ_i equals 0. In fact, suppose $du_i = \sum_{j\neq i} a_j\, du_j + \sum_j b_j\, dv_j$, where the a_j, b_j are real numbers. Then $\{u_i, v_i\} = \sum_{j\neq i} a_j \{u_j, v_i\} + \sum_j b_j \{v_j, v_i\}$. By (3.8) the right-hand side equals 0, and so $\{u_i, v_i\} = 0$. But this contradicts the fact that $\{u_i, v_i\} = -\frac{\lambda_i}{2}$ with $\lambda_i \neq 0$. We argue exactly the same way to show that dv_i is not a linear combination of the other differentials. Since the du_i and dv_i are linearly independent at (x_o, ξ_o) (thus showing that Σ is a manifold near (x_o, ξ_o)) and so form a basis for $N(\Sigma)_{(x_o,\xi_o)}$, it follows from (3.8) and the fact that no λ_i equals 0 that $\omega_{(x_o,\xi_o)}$ on $N(\Sigma)_{(x_o,\xi_o)}$ is non-degenerate.

We now want to analyze in some detail the Laplacians $\Delta_k = P\,P^* + P^*\,P : E^k \longrightarrow E^k$. We assume that the Hermitian metrics in the bundles $E^k = E^o \otimes \Lambda^k W$ have been chosen as follows. We give E^o and W arbitrary Hermitian metrics, then give $\Lambda^k W$ the metric induced by that on W , and finally give E^k the "product metric". The distinguished basis $w_1,\dots,w_q$ for W is taken to be orthonormal, and so $w_{i_1}\wedge\cdots\wedge w_{i_k}$, $i_1<\cdots<i_k$ form an orthonormal basis for $\Lambda^k W$. (Notice that in diagonalizing the Levi-form at the fixed point (x_o,ξ_o) by means of a unitary transformation we need not disturb the metrics in the E^k ; indeed we simply let the corresponding unitary transformation act on the w_i 's and thus get a new orthonormal basis for W.)

Now let w_I, w_J denote $w_{i_1}\wedge\cdots\wedge w_{i_k}$, $w_{j_1}\wedge\cdots\wedge w_{j_k}$, respectively. Then, letting $\langle\ ,\rangle$ denote Hermitian metric we have that

$$(3.9)\qquad \langle w_I, w_J\rangle = \delta_{IJ} \quad .$$

Since $P\,(f_I \otimes w_I) = \sum_{i=1}^{q} P_i\,(f_I) \otimes (w_i \wedge w_I)$ it follows from (3.9) that $P^* : E^k \longrightarrow E^{k-1}$ is given by

$$(3.10)\qquad P^*\,(f_I \otimes w_I) = \sum_{i=1}^{q} P_i^*\,(f_I) \otimes (w_i \lrcorner\, w_I) \quad .$$

Therefore, $\Delta_k (f_I \otimes w_I) \equiv P P^* (f_I \otimes w_I) + P^* P (f_I \otimes w_I)$

$$= \sum_{i,j}^{q} (P_i P_j^* f_I) \otimes (w_i \wedge (w_j \lrcorner w_I))$$

$$+ \sum_{i,j}^{q} (P_j^* P_i f_I) \otimes (w_j \lrcorner (w_i \wedge w_I)) \quad .$$

Writing $P_i P^*_j = [P_i, P_j^*] + P_j^* P_i$ we can rewrite the preceding as $\Delta_k (f_I \otimes w_I) = \sum_{i,j} (P_j^* P_i f_I) \otimes [(w_i \wedge (w_j \lrcorner w_I)) +$

$$(w_j \lrcorner (w_i \wedge w_I))] + \sum_{i,j} [P_i, P_j^*] (f_I) \otimes$$

$$(w_i \wedge (w_j \lrcorner w_I))$$

Since the w_i are orthogonal we have

$$(3.11) \qquad w_i \wedge (w_j \lrcorner w_I) + w_j \lrcorner (w_i \wedge w_I) = \begin{cases} w_I & \text{if } i = j \\ 0 & \text{if } i \neq j \end{cases} \quad .$$

Hence, the preceding can be rewritten as

$$(3.12) \qquad \Delta_k (f_I \otimes w_I) = (\sum_{i=1}^{q} P_i^* P_i) f_I \otimes w_I +$$

$$\sum_{i,j} [P_i, P_j^*] (f_I) \otimes (w_i \wedge (w_j \lrcorner w_I)) \quad .$$

We next determine $(\tilde{\Delta}_k)_{(x_o, \xi_o)}$ and $\sigma_{sub} (\Delta_k)|_{(x_o, \xi_o)}$. It is clear from (3.12) that the principal symbol , $\sigma_2 (\Delta_k)$, of the second-order operator Δ_k is given by

$$(3.13)\qquad \sigma_2(\Delta_k) = \sum_{i=1}^{q} \bar{p}_i \, p_i = \sum_{i=1}^{q} (u_i^2 + v_i^2) \quad .$$

But now, using the commutation relations (3.8) and the fact that $\lambda_i > 0$, $i = 1,\ldots,r$, $\lambda_i < 0$, $i = r+1,\ldots,q$, we see that if we define

$$(3.14)\qquad u_i' = \begin{cases} \sqrt{\dfrac{2}{\lambda_i}}\, u_i & i = 1,\ldots,r \\ -\sqrt{\dfrac{-2}{\lambda}}\, u_i & i = r+1,\ldots,q \end{cases}$$

$$v_i' = \begin{cases} \sqrt{\dfrac{2}{\lambda_i}}\, v_i & i = 1,\ldots,r \\ \sqrt{\dfrac{-2}{\lambda_i}}\, v_i & i = r+1,\ldots,q \end{cases}$$

then the u_i' 's and v_i' 's satisfy the canonical commutation relations

$$(3.15)\qquad \{u_i', u_j'\}|_{(x_0,\xi_0)} = 0 \quad , \quad \{v_i', v_j'\}|_{(x_0,\xi_0)} = 0 \; ,$$

$$\{u_i', v_j'\}|_{(x_0,\xi_0)} = -\delta_{ij} \quad .$$

In these canonical coordinates $\sigma_2(\Delta_k)$ has the expression

$$(3.16)\qquad \sigma_2(\Delta_k) = \sum_{i=1}^{r} \frac{\lambda_i}{2} (u_i'^2 + v_i'^2) + \sum_{i=r+1}^{q} \frac{(-\lambda_i)}{2} (u_i'^2 + v_i'^2) \quad .$$

Thus we know (see, for example, (2.37) - (2.43)) that $(\tilde{\Delta}_k)_{(x_o,\xi_o)}$ is given by

$$(3.17)\qquad (\tilde{\Delta}_k)_{(x_o,\xi_o)} = \sum_{i=1}^{r} \frac{\lambda i}{2} (D_{s_i}^2 + s_i^{\,2}) + \sum_{i=r+1}^{q} - \frac{\lambda i}{2} (D_{s_i}^2 + s_i^{\,2}) \quad .$$

Hence, as we know from §2.5, the eigenvalues of $(\tilde{\Delta}_k)_{(x_o,\xi_o)}$ are precisely all numbers of the form $\sum_{i=1}^{r} \frac{\lambda i}{2}(2n_i + 1) + \sum_{i=r+1}^{q} - \frac{\lambda i}{2}(2n_i + 1)$, where $n_1,\ldots,n_q$ are any <u>non-negative</u> integers. It is convenient to rewrite these numbers in the form

$$(3.18)\qquad \sum_{i=1}^{r} \lambda_i n_i - \sum_{i=r+1}^{q} \lambda_i n_i + \sum_{i=1}^{r} \frac{\lambda i}{2} + \sum_{i=r+1}^{q} \left(\frac{-\lambda i}{2}\right)$$

Next we compute $\sigma_{sub}(\Delta_k)|_{(x_o,\xi_o)}$. Recall (§2.2, Prop 2.19) that $\sigma_{sub}(\Delta_k)|_{(x_o,\xi_o)} = \sigma_1(\Delta_k)|_{(x_o,\xi_o)} - \frac{1}{2\sqrt{-1}} \sum_{\ell} \frac{\partial^2}{\partial\xi_\ell \partial x_\ell} \sigma_2(\Delta_k)|_{(x_o,\xi_o)}$.

Using the symbol calculus for pseudo-differential operators we see that

$$\sigma_1 (P_i^* P_i)|_{(x_o,\xi_o)} = \sigma_1 (P_i^*)\, \sigma_o (P_i)|_{(x_o,\xi_o)} + \sigma_o (P_i^*)\, \sigma_1 (P_i)|_{(x_o,\xi_o)} + \frac{1}{\sqrt{-1}} \sum_{\ell} \frac{\partial}{\partial\xi_\ell} \sigma_1(P_i^*) \frac{\partial}{\partial x_\ell} \sigma_1(P_i)|_{(x_o,\xi_o)}$$

Since $\sigma_1(P_i) = p_i$ and $\sigma_1(P_i^*) = \overline{p}_i$, and since (x_o, ξ_o) is in Σ , it follows that the first two terms vanish, so that

$$(3.19) \qquad \sigma_1 (P_i^* P_i)|_{(x_o,\xi_o)} = \frac{1}{\sqrt{-1}} \sum_{\ell} \frac{\partial \overline{p}_i}{\partial \xi_\ell} \frac{\partial p_i}{\partial x_\ell} \Big|_{(x_o,\xi_o)}$$

Since $\sigma_1 ([P_i, P_j^*])|_{(x_o,\xi_o)} = \frac{1}{\sqrt{-1}} \{p_i, \overline{p}_j\}|_{(x_o,\xi_o)}$ $= \lambda_i \delta_{ij}$, it follows from (3.12) and (3.19) that $\sigma_1(\Delta_k)|_{(x_o,\xi_o)} : E^o \otimes \Lambda^k W \longrightarrow E^o \otimes \Lambda^k W$ is the matrix given by

$$(3.20) \qquad \sigma_1(\Delta_k)|_{(x_o,\xi_o)} : e_I \otimes w_I \longmapsto$$

$$\sum_{i=1}^{q} \frac{1}{\sqrt{-1}} \sum_{\ell} \frac{\partial \overline{p}_i}{\partial \xi_\ell} \frac{\partial p_i}{\partial x_\ell} \Big|_{(x_o,\xi_o)} e_I \otimes w_I$$

$$+ \sum_{i=1}^{q} \lambda_i \, e_I \otimes (w_i \wedge (w_i \lrcorner w_I)) \ .$$

But now we examine $\frac{1}{2\sqrt{-1}} \sum_{\ell} \frac{\partial^2}{\partial \xi_\ell \partial x_\ell} \sigma_2(\Delta_k)\Big|_{(x_o,\xi_o)}$.

From (3.13) and the fact that $(x_o, \xi_o) \in \Sigma$ it follows that

$$(3.21) \qquad \frac{1}{2\sqrt{-1}} \sum_{\ell} \frac{\partial^2}{\partial \xi_\ell \partial x_\ell} \sigma_2(\Delta_k)|_{(x_o,\xi_o)} : e_I \otimes w_I \longmapsto$$

$$\sum_{i=1}^{q} \frac{1}{2\sqrt{-1}} \sum_{\ell} \left(\frac{\partial \overline{p}_i}{\partial \xi_\ell} \frac{\partial p_i}{\partial x_\ell} + \frac{\partial \overline{p}_i}{\partial x_\ell} \frac{\partial p_i}{\partial \xi_\ell} \right)\Big|_{(x_o,\xi_o)} e_I \otimes w_I \ .$$

From the definition of $\sigma_{sub}(\Delta_k)$ and from (3.20), (3.21) we have

$$(3.22) \qquad \sigma_{sub}(\Delta_k)|_{(x_o,\xi_o)} : e_I \otimes w_I \longmapsto$$

$$\sum_{i=1}^{q} \tfrac{1}{2}\{\overline{p}_i, p_i\}|_{(x_o,\xi_o)} \; e_I \otimes w_I + \sum_{i=1}^{q} \lambda_i \, e_I \otimes$$

$$(w_i \wedge (w_i \lrcorner\, w_I)) = \Big(\sum_{i=1}^{q} - \frac{\lambda_i}{2}\Big) e_I \otimes w_I + \sum_{i=1}^{q} \lambda_i \, e_I \otimes$$

$$(w_i \wedge (w_i \lrcorner\, w_I)) .$$

But observe that $w_i \wedge (w_i \lrcorner\, w_I) = \begin{cases} w_I & \text{if } i \in \{i_1,\ldots,i_k\}, \\ & \text{where } w_I = w_{i_1}\wedge\cdots\wedge w_{i_k} \\ 0 & \text{otherwise} \end{cases}$.

Hence

$$(3.33) \qquad \sigma_{sub}(\Delta_k)|_{(x_o,\xi_o)} : e \otimes w_{i_1} \wedge\cdots\wedge w_{i_k} \longmapsto$$

$$\Big[\sum_{i=1}^{q} \Big(-\frac{\lambda_i}{2}\Big) + \sum_{n=1}^{k} \lambda_{i_n}\Big] e \otimes w_{i_1}\wedge\cdots\wedge w_{i_k}$$

Since the elements of the form $w_{i_1}\wedge\cdots\wedge w_{i_k}$, $i_1<\cdots<i_k$, form a basis for $\Lambda^k W$ it follows from (3.33) that the eigenvalues of the matrix $\sigma_{sub}(\Delta_k)|_{(x_o,\xi_o)}$ are precisely all numbers of the form $\sum_{i=1}^{q} \left(-\frac{\lambda_i}{2}\right) + \sum_{n=1}^{k} \lambda_{i_n}$, where the i_n's are k distinct integers between 1 and q, inclusive.

But Theorem 2.4 tells us that Δ_k is hypoelliptic at (x_o, ξ_o) with loss of one derivative if and only if no eigenvalue of the matrix $\sigma_{sub}(\Delta_k)|_{(x_o,\xi_o)}$ equals -1 times an eigenvalue of the operator $(\tilde{\Delta}_k)_{(x_o,\xi_o)}$. That is, hypoellipticity with loss of one derivative is equivalent to the following equality (3.34) <u>not</u> holding for any choice of <u>non-negative</u> integers n_i and any choice of k distinct integers i_n between 1 and q.

$$(3.34) \quad \sum_{i=1}^{r} \lambda_i n_i - \sum_{i=r+1}^{q} \lambda_i n_i + \sum_{i=1}^{r} \frac{\lambda_i}{2} + \sum_{i=r+1}^{q} \left(-\frac{\lambda_i}{2}\right) = -\left[\sum_{i=1}^{r} \left(-\frac{\lambda_i}{2}\right) + \sum_{i=r+1}^{q} \left(-\frac{\lambda_i}{2}\right) + \sum_{n=1}^{k} \lambda_{i_n}\right].$$

We can rewrite (3.34) as

$$(3.35) \quad \sum_{n=1}^{k} \lambda_{i_n} = \sum_{i=r+1}^{q} \lambda_i(n_i + 1) - \sum_{i=1}^{r} \lambda_i n_i .$$

We shall see that (3.35) holds for some choice of <u>non-negative</u> integers n_i and some choice of k distinct integers i_n between 1 and q precisely if $k = q - r$, i.e., if k equals the number of negative eigenvalues of the Levi-form. Thus, Δ_k is hypoelliptic at (x_o, ξ_o) with loss of one derivative precisely if $k \neq q - r$. Of course, this condition is equivalent to the condition $r \geqq (q - k+1)$ or $k + 1 \leqq q - r$, which states that the Levi-form has at least $(q - k+1)$ positive or at least $(k+1)$ negative eigenvalues. Since the Levi-form is non-degenerate, i.e. , since no eigenvalue equals 0 , this criterion agrees with the criterion (see, for example, [15], [22], [23]) for $\frac{1}{2}$-subellipticity at (x_o, ξ_o) of the complex $\{P, E^i\}$ at position E^k .

The proof of the above assertion is very easy. If $k > q-r$, then there is at least one positive eigenvalue among the $\{\lambda_{i_n}\}$. But since $n_i + 1 \geqq 1$ and since all the negative eigenvalues occur among the λ_i , $i = r+1,\ldots,q$ it follows that $\sum_{n=1}^{k} \lambda_{i_n} > \sum_{i=r+1}^{q} \lambda_i(n_i + 1)$. Since $\lambda_i > 0$ for $i = 1,\ldots,r$, $-\sum_{i=1}^{r} \lambda_i n_i < 0$. Taken together with the preceding inequality this yields

$$\sum_{n=1}^{k} \lambda_{i_n} > \sum_{i=r+1}^{q} \lambda_i(n_i + 1) - \sum_{i=1}^{r} \lambda_i n_i$$

Thus, (3.35) fails to hold.

If $k < q - r$ then at least one of the negative eigenvalues does not appear in the sum $\sum_{n=1}^{k} \lambda_{i_n}$. Hence, again using the fact that $n_i + 1 \geqq 1$ and the fact that all the negative eigenvalues occur among the λ_i, $i = r+1,\ldots,q$, we again see that $\sum_{n=1}^{k} \lambda_{i_n} > \sum_{i=r+1}^{q} \lambda_i (n_i + 1)$. Again, since

$- \sum_{i=1}^{r} \lambda_i n_i \leqq 0$ (the equality holding if $r = 0$) it follows that $\sum_{n=1}^{k} \lambda_{i_m} > \sum_{i=r+1}^{q} \lambda_i (n_i + 1) - \sum_{i=1}^{r} \lambda_i n_i$. Thus, again, (3.35) fails to hold.

Finally, suppose $k = q - r$. Then there are precisely k negative λ_i 's . Taking these as our λ_{i_n} , $n = 1,\ldots,k$, and taking all the n_i equal to 0 , we see that (3.35) holds.

§4. Hypoellipticity and asymptotic eigenvalues in the abstract case

In this section we shall show how the results and methods of Treves ([25]) may be interpreted from the stand-point of asymptotic eigenvalues. The eigenvalues of the test-operators in §2 will appear, in this context, as the leading terms in the asymptotic expansions of the operators occurring here. It is mainly the general view-point which we wish to emphasize. Indeed, there is essentially no technical contribution here to the work of [25]. For this reason, and also because there is so great a resemblance to the eigenvalue computation of §2, we shall be very sketchy in presenting details.

The operators treated in [25] are abstract second-order evolution operators of the type

$$(4.1) \qquad P = \left(\frac{\partial}{\partial t} - a(t,A)A\right)\left(\frac{\partial}{\partial t} - b(t,A)A\right) - c(t,A)A \quad .$$

Here A is an <u>unbounded</u>, densely defined, self-adjoint positive-definite linear operator on a Hilbert space H, with <u>bounded</u> inverse A^{-1}. The expressions $a(t,A)$, $b(t,A)$, $c(t,A)$ are power series in non-negative powers of A^{-1} with coefficients C^∞ functions of t. These power series, as well as all their t-derivatives are assumed to converge uniformly on compact sets (in $\mathbb{R}$) in

$B(H, H)$, the space of bounded linear operators on H . Treves makes the restriction that the real parts of $a_o(t)$ and $b_o(t)$, the leading coefficients of $a(t,A)$, $b(t,A)$, respectively, vanish at $t = 0$, but that $\operatorname{Re} a_o'(0) \neq 0$ and $\operatorname{Re} b_o'(0) \neq 0$.

Let H^N denote the domain of A^N for $N \geqq 0$, and let $H^\infty = \bigcap_N H^N$. Let H^{-N} be the dual space of H^N , and $H^{-\infty}$ the dual space of H^∞ . (Of course, if $H = L^2(\mathbb{R}^n)$ and if $A = \Lambda$, the square-root of the Laplacian, then the above H^N spaces are just the usual Sobolev spaces.) In this context one defines, given an interval J in $\mathbb{R}$ containing 0 , an appropriate space of C^∞ functions, $C^\infty(J, H^\infty)$, and a space of distributions $\mathcal{D}'(J, H^{-\infty})$. The operator P may be viewed both as a map from $\mathcal{D}'(J, H^{-\infty})$ into $\mathcal{D}'(J, H^{-\infty})$ and as a map from $C^\infty(J, H^\infty)$ into $C^\infty(J, H^\infty)$. In this context one makes the obvious definitions of hypoellipticity and local solvability. We shall deal only with the hypoellipticity definition.

Definition 4.1. P is hypoelliptic in J if, for any open subset J' of J and any distribution $u \in \mathcal{D}'(J', H^{-\infty})$

$$Pu \in C^\infty(J', H^{-\infty}) \Longrightarrow u \in C^\infty(J', H^\infty) . \tag{4.1}$$

P is hypoelliptic at a point t_o if there exists an open interval J containing t_o such that P is hypoelliptic in J .

Treves shows that if $\operatorname{Re} a_o'(0) > 0$ and $\operatorname{Re} b_o'(0) > 0$ then P is hypoelliptic but not locally solvable at $t = 0$; that if $\operatorname{Re} a_o'(0) < 0$ and $\operatorname{Re} b_o'(P) < 0$ then P is locally solvable but not hypoelliptic at $t = 0$; and that if $\operatorname{Re} a_o'(0)$ and $\operatorname{Re} b_o'(0)$ have opposite sign then P is locally solvable at $t = 0$ if and only if P is hypoelliptic at $t = 0$, and, moreover, these equivalent conditions hold precisely when no $c^j(A)A$, $j = 0,1,\ldots$ vanishes identically. Here $c^j(A)$, $j = 0,1,\ldots$ is a sequence of formal (i.e., not necessarily convergent) power series in non-negative powers of A^{-1} with coefficients in $\mathbb{C}$. Treves derives the formal power series $c^j(A)A$ by means of his general method of concatenations. We shall see that $-c^j(A)A$, $j = 0,1,\ldots$ are the "asymptotic eigenvalues" of P , and that the method of concatenations is a generalization of the procedure in §2.5 for computing eigenvalues by means of commutation relations and a boundedness below condition. (See also §2.4 for a discussion of "conflicting influences".)

We recall how the power series $c^j(A)A$ arise. Let

$$(4.2) \qquad X = \frac{\partial}{\partial t} - a(t,A)A \quad \text{and} \quad Y = \frac{\partial}{\partial t} - b(t,A)A \ .$$

Treves shows how to generate from the original operator P a sequence of formal operators.

$$(4.3)\quad \begin{cases} P^j = X^j\, Y^j - c^j(A)A & j = 0,1,\ldots \\ X^j = \partial_t - \varphi^j(t,A)A & \\ Y^j = \partial_t - \psi^j(t,A)A & . \end{cases}$$

By <u>formal</u> we mean that the coefficients are <u>formal</u> power series of the form

$$(4.4)\quad \begin{cases} c^j(A) = \sum_{i=0}^{\infty} c_i^j\, A^{-i} & , \; c_i^j \in \mathbb{C} \\ \varphi^j(t,A) = \sum_{i=0}^{\infty} \varphi_i^j(t)\, A^{-i} & , \; \varphi_i^j \in C^\infty(J) \\ \psi^j(t,A) = \sum_{i=0}^{\infty} \psi_i^j\,(t)\, A^{-i} & , \; \psi_i^j \in C^\infty(J) \quad . \end{cases}$$

These operators satisfy the following properties:

(4.5) $\qquad P^o$ is formally equal to P .

$$(4.6)\qquad \varphi_o^j(t) = a_o(t) \;\; , \; \psi_o^j(t) = b_o(t) \;\; , \text{ for every } j \; .$$

$$(4.7)\qquad c_o^j = c_o(0) + j\, \delta_o'(0) \qquad , \text{ for every } j \; .$$

Here $a_o(t)$, $b_o(t)$, $c_o(t)$ are the leading coefficients of $a(t,A)$, $b(t,A)$, $c(t,A)$, respectively, and $\delta_o(t) = a_o(t) - b_o(t)$. (Notice that under the hypothesis

(4.8) $\quad \mathrm{Re}\, a_o'(0) > 0 \quad , \quad \mathrm{Re}\, b_o'(0) < 0$

it follows, in particular, that $\delta_o'(0) \neq 0$.)

Moreover, the following commutation relations are satisfied.

(4.9) $\quad X^j P^{j+1} = P^j X^j \quad$, for every j .

(4.10) $\quad Y^j P^j = P^{j+1} Y^j \quad$, for every j .

We need the following definition.

Definition 4.2. Let $f(x)$ be a (bounded) complex-valued C^∞ function on $\mathbb{R}^+$. We say that the formal power series $\sum_{r=0}^{\infty} a_r A^{-r}$ represents $f(A)$ asymptotically (and write $f(A) \sim \sum_{r=0}^{\infty} a_r A^{-r}$) if and only if for every $m \geqq 0$

(4.11) $\quad f(A) - \sum_{r=0}^{m} a_r A^{-r}$ maps H into H^{m+1}

and

(4.12) $\quad A^{m+1} (f(A) - \sum_{r=0}^{m} a_r A^{-r})$ as an operator from H to H is bounded.

Remarks: 1. Not every bounded function f of A need

have an asymptotic expansion, but the asymptotic expansion of $f(A)$, if it exists, is unique. This follows easily from (4.11) and (4.12) and the fact that A is unbounded.

2. Every formal power series $\sum_{r=0}^{\infty} a_r A^{-r}$ is the asymptotic expansion of some $f(A)$. This follows immediately from the corresponding fact about the formal power series $\sum_{r=0}^{\infty} a_r x^{-r}$ (i.e., there exists an (analytic) function $f(x) : R^+ \longrightarrow ¢$ such that $f(x) \sim \sum_{r=0}^{\infty} a_r x^{-r}$) and the spectral theorem.

We shall write $f(A) \sim 0$ if $a_r = 0$ for every r.

Observe that (4.8) implies, in particular, that $\operatorname{Re} a_0(t) \neq 0$, $\operatorname{Re} b_0(t) \neq 0$ for $t \neq 0$ and t sufficiently close to 0, say in the interal J. Thus P is "elliptic" (e.g., if $A = |D_x|$), and so, in particular, hypoelliptic for $t \neq 0$. Hence the condition for hypoellipticity at $t = 0$ can be stated as follows.

(4.13) For every $\psi \in \mathcal{D}'(J, H^{-\infty})$, if there exists an interval $\{0\} \subset J' \subset J$ such that $P\psi \in C^{\infty}(J', H^{\infty})$, then $\psi \in C^{\infty}(J'', H^{\infty})$ for some interval $\{0\} \subset J'' \subset J$.

It will be slightly more convenient to work with this formulation rather than with the definition itself, but we could use the latter if we wished. [Note that, because of (4.6), we can use the same "ellipticity" interval J for every P^j. We tacitly use this fact in the proof of Theorem 4.5 below.]

Definition 4.3. Let $\psi \in \mathcal{D}'(J, H^{-\infty})$. We write $\psi \cong 0$ if and only if there exists an interval $\{0\} \subset J^1 \subset J$ such that $\psi \in C^\infty(J', H^\infty)$. Clearly, for any operator P of type (4.1) $\psi \cong 0 \Longrightarrow P\psi \cong 0$.

Definition 4.4. Let $f(A) \sim \sum_{r=0}^{\infty} a_r A^{-r}$. Then we say that $f(A)A$ is an "eigenvalue" of P (and $(\sum_{r=0}^{\infty} a_r A^{-r})A$ is an asymptotic eigenvalue of P) if and only if there exists $\psi \in D'(J, H^{-\infty})$ such that $\psi \not\cong 0$ but such that $(P - f(A)A)\psi \cong 0$.

Of course, in view of (4.13), $f(A)A$ is an "eigenvalue" of P precisely if $P - f(A)A$ is not hypoelliptic, and in particular P is hypoelliptic precisely if 0 is not an eigenvalue of P. Treves shows, in effect, that this is equivalent to 0 not being an asymptotic eigenvalue of P, and makes this latter condition more explicit by determining precisely what are the asymptotic eigenvalues of P.

Theorem 4.5. The formal power series $-c^j(A)A$, $j = 0,1,\ldots$ are precisely the asymptotic eigenvalues of P . Thus, the hypoellipticity criterion for P is that no $c^j(A)A$ is identically 0 . As in §2.5, because of "conflicting influences" this is also the local solvability criterion for P .

Before proceeding to the proof of Theorem 4.5 we note that, after verifying that Treves' arguments work just as well for asymptotic expansions as convergent expansions, one would obtain Theorem 4.5 as an immediate consequence of Treves' results. However, we give a sketch of the proof, not differing in any essential way from that in [25] , but stressing the similarity with the computation of §2.5, and presenting the method of "concatenations" as a procedure for computing asymptotic eigenvalues, generalizing the computation of the eigenvalues of the harmonic oscillator. Of course, this is not to minimize the extent to which the method of "concatenations" generalizes the classical procedure.

Pf:

Assume first that we are in the "convergent" case, so that the "formal" operators P^j are genuine operators. We begin with a boundedness below condition on the eigenvalues. Indeed, Treves' subelliptic estimate

(Theorem II. 2.1 of [25] ; see also Cor II. 3.2) shows that under the hypotheses (4.6), (4.8) there is a <u>uniform</u> (i.e., independent of j) lower bound on the leading terms of the eigenvalues of $N^j \equiv X^j Y^j$. More precisely, the following holds.

(4.14) For any j , if $f(A)A \sim (\sum_{r=0}^{\infty} a_r A^{-r})A$ is an eigenvalue of N^j , then $\mathrm{Re}(-\bar{\delta}_o'(0)\, a_o) > -\frac{1}{2}|\delta_o'(0)|^2$.

(Compare §2.5, (2.133b).)

We need some further preliminaries (proved in [25]). Let $d(t,A)$ be a (convergent) power series of the type $\sum_{i=0}^{\infty} d_i(t) A^{-i}$. Then if $\mathrm{Re}\, d_o(0) \neq 0$, then $\partial_t - d(t,A)A$ is "elliptic" at $t = 0$, and in particular hypoelliptic and locally solvable. If $\mathrm{Re}\, d_o(0) = 0$ and $\mathrm{Re}\, d_o'(0) > 0$, then $\partial_t - d(t,A)A$ is hypoelliptic at $t = 0$ but not locally solvable, and if $\mathrm{Re}\, d_o(0) = 0$ and $\mathrm{Re}\, d_o'(0) < 0$, then $\partial_t - d(t,A)A$ is locally solvable at $t = 0$ but not hypoelliptic. In particular

(4.15) X^j is hypoelliptic, Y^j is not hypoelliptic, $j = 0,1,\ldots$

We can now proceed along the same general lines as in

§2.5. First, using (4.10) we show that if $f(A)A \sim (\sum_{r=0}^{\infty} a_r A^{-r})A$ is an eigenvalue of N^o then $\sum_{r=0}^{\infty} a_r A^{-r} = c^o(A) - c^j(A)$ for some $j = 0,1,\ldots$. Letting $\lambda^{j+1}(A) \equiv c^{j+1}(A) - c^j(A)$, $j = 0,1,\ldots$, we rewrite (4.10) as

$$(4.16) \qquad Y^j N^j = (N^{j+1} - \lambda^{j+1}(A)A) Y^j \quad , \; j = 0,1,\ldots$$

Suppose $f(A)A$ is an eigenvalue of N^o . Then there exists ψ such that $\psi \not\cong 0$ but such that $(N^o - f(A)A)\psi \cong 0$. Applying (4.16) we get

$$[N^1 - (\lambda^1(A)A + f(A)A)] Y^o \psi \cong 0 \ .$$

So, either $Y^o\psi \cong 0$ or $\lambda^1(A)A + f(A)A$ is an eigenvalue of N^1 . Suppose $Y^o \psi \cong 0$. Then since $N^o = X^o Y^o$, $N^o\psi \cong 0$. But, by assumption, $N^o \psi \cong f(A)A \psi$. Thus, $f(A)A\psi \cong 0$. Following §2.5 , we want to show that, since $\psi \not\cong 0$, $f(A)A \sim 0$. This requires a bit more work than in §2.5. One shows

$$(4.17) \qquad \psi \not\cong 0 \ , \ Y^o\psi \cong 0 \ , \text{ and } \ f(A)A \psi \cong 0 \Longrightarrow f(A)A \sim 0 \quad .$$

The argument is simply that if $f(A)A \not\sim 0$, then, taking the smallest j such that $a_j \neq 0$, we get that $Y^o - \overline{a}_j f(A)A^{j+1}$ is "elliptic" in some neighborhood containing 0. But $Y^o\psi \cong 0$ and $f(A)A\psi \cong 0 \Longrightarrow$ $(Y^o - \bar{a}_j f(A)A^{j+1})\psi \cong 0 \Longrightarrow$(by "ellipticity") $\psi \cong 0$. This proves (4.17).

We have shown (writing $\lambda^o(A)A \equiv 0$)

(4.18) Either $f(A)A + \lambda^o(A)A \sim 0$ or $f(A)A + \sum_{i=0}^{1} \lambda^i(A)$ is an eigenvalue of N^1.

If $f(A)A + \lambda^o(A)A \not\sim 0$ we can then repeat the above argument, with N^1 replacing N^o, N^2 replacing N^1, and $f(A)A + \sum_{i=0}^{1} \lambda^i(A)A$ replacing $f(A)A$, and obtain that either $f(A)A + \sum_{i=0}^{1} \lambda^i(A)A \sim 0$ or $f(A)A + \sum_{i=0}^{2} \lambda^i(A)A$ is an eigenvalue of N^2. Iterating this process we see that if for every $k < j$ $f(A)A + \sum_{i=0}^{k} \lambda^i(A)A \not\sim 0$, then

(4.19) Either $f(A)A + \sum_{i=0}^{j} \lambda^i(A)A \sim 0$ or $f(A)A + \sum_{i=0}^{j+1} \lambda^i(A)A$ is an eigenvalue of N^{j+1}.

But $\sum_{i=0}^{j+1} \lambda^i(A)A = c^{j+1}(A)A - c^o(A)A$, and so by (4.7) the leading term of $f(A)A + \sum_{i=0}^{j+1} \lambda^i(A)A$ is $j\,\delta_o'(0) + a_o$.

Hence, $\mathrm{Re}(-\bar{\delta}_o'(0) \cdot \text{leading term}) = -j|\delta_o'(0)|^2 + \mathrm{Re}\,\bar{\delta}_o'(0)\,a_o$. Clearly, for j sufficiently large this violates the lower bound of (4.14). Thus, just as in

§2.5, we see that, for some j, $f(A)A \sim c^{o}(A)A - c^{j}(A)A$. This shows that any asymptotic eigenvalue of P (which in the convergent case equals P^{o}) is $-c^{j}(A)A$ for some j.

To show that, for every j, $-\sum_{i=0}^{j} \lambda^{i}(A)A$ is an (asymptotic) eigenvalue of N^{o} (and hence that, for every j, $-c^{j}(A)A$ is an asymptotic eigenvalue of P) we use (4.9) in the form

$$X^{j}\, N^{j+1} = (N^{j} + \lambda^{j+1}(A)A)\, X^{j} \quad . \tag{4.17}$$

Since X^{j} is hypoelliptic (and so $X^{j}\,\psi \cong 0 \Longrightarrow \psi \cong 0$), (4.17) shows that if $\mu(A)A$ is an eigenvalue of N^{j+1}, then $\mu(A)A - \lambda^{j+1}(A)A$ is an eigenvalue of N^{j}. Iterating, we see that if $\mu(A)A$ is an eigenvalue of N^{j} then $\mu(A)A - \sum_{i=0}^{j} \lambda^{i}(A)A$ is an eigenvalue of N^{o}. Thus, it suffices to show that, for every j, 0 is an eigenvalue of N^{j}. But since Y^{j} is not hypoelliptic there exists ψ such that $\psi \not\cong 0$ but such that $Y^{j}\,\psi \cong 0$. Hence $N^{j}\psi = X^{j}\, Y^{j}\psi \cong 0$. This completes the argument in the convergent case.

The non-convergent case is essentially similar, although quite a bit more delicate technically. One works with truncated power series, and instead of the equivalence relation $\cong$, one uses much sharper L^{2} - statements. To prove that every asymptotic eigenvalue of P is of the form $-c^{j}(A)A$ one again uses

(4.10) , this time with a sharper boundedness below condition ([25], Propl II. 3.3) replacing (4.14) . To show that every $-c^j(A)A$ is in fact an asymptotic eigenvalue of P one can use (4.9) and the estimate ([25], Prop I. 1.2), as in an earlier, unpublished version of [25], or proceed as in the published version of [25], which uses (4.9), but combines the treatment of hypoellipticity and local solvability. We omit the details.

In conclusion, we observe that if we form the "test-operator" for hypoellipticity with loss of 1 derivative for P at the point $t = 0$, with A viewed as ξ , or better, as $\rho\xi_0$, then it is clear from the above computation that the exact eigenvalues of the test-operator are precisely $-c_0^j(A)A$, $j = 0,1,\ldots,$ i.e. ,the leading terms of the asymptotic eigenvalues $-c^j(A)A$ of the full operator P . So the "strongest" hypoellipticity condition states that no eigenvalue of P has its leading term equal to 0 , whereas the general hypoellipticity condition insists only that no eigenvalue has an identically 0 asymptotic expansion. It may be of interest to study the sequence of successively weaker hypoellipticity conditions which thus arises, the k-th condition stating that no asymptotic eigenvalue $-c^j(A)A$ has its first k terms all equal to 0 , i.e., that no $-c^j(A)A$ is $O(A^{-(k-1)})$. Related to this is the question of whether it is possible to construct more refined "test operators" whose exact eigenvalues (in the classical sense) would correspond to the first k terms of the $-c^j(A)A$.

Appendix: Remark on simple characteristics

If P has principal symbol $p = u + iv$ vanishing only to first order on Σ, assumed non-involutive, i.e., $\{u,v\} \neq 0$, then although $\sigma_{sub}(P)|_{\Sigma}$ is not defined, there is still a reasonable candidate for "test-operator" at $(x,\xi) \in \Sigma$. In fact, in analogy with Remark 2.11, define $\tilde{p}_{(x,\xi)} : N(\Sigma)_{(x,\xi)} \longrightarrow \mathbb{C}$ by $\tilde{p}_{(x,\xi)}(df_{(x,\xi)}) = H_f(p)_{(x,\xi)} = \{f,p\}_{(x,\xi)}$. Then, as in §2.2 we can associate to $\tilde{p}_{(x,\xi)}$ an operator $\tilde{P}_{(x,\xi)}$, defined up to unitary equivalence. To see that $\tilde{P}_{(x,\xi)}$ is a reasonable candidate for "test-operator", we proceed as follows. Assume for simplicity, (and without loss of generality) that $\{u,v\}_{(x,\xi)} = -1$ or $\{u,v\}_{(x,\xi)} = 1$. In the first case, u, v satisfy the canonical commutation relations and so if we use coordinates s, t on $N(\Sigma)_{(x,\xi)}$ given by $(s,t) \longmapsto sdu + tdv$, then $\omega_{(x,\xi)} = dt \wedge ds$. Hence, $\tilde{p}_{(x,\xi)}(s,t) \equiv \tilde{p}_{(x,\xi)}(sdu + tdv) = \{su + tv, p\} = \{su + tv, u + iv\} = is\{u,v\} + t\{v,u\} = -is + t$. Thus, if $\{u,v\}_{(x,\xi)} = -1$, $\tilde{P}_{(x,\xi)} = D_s - is$. Similarly, if $\{u,v\}_{(x,\xi)} = 1$ we can use u, $-v$ as canonical coordinates, and we see that $\tilde{P}_{(x,\xi)} = D_x + is$. We know that for any $\lambda \in \mathbb{C}$ $D_s - is + \lambda$, as an operator from $H_{(1,1)}(\mathbb{R}^1)$ to $L^2(\mathbb{R}^1)$ is surjective but not injective, whereas $D_x + is + \lambda$ is injective but not surjective. But $\{u,v\}_{(x,\xi)} = -1$, $\{u,v\}_{(x,\xi)} = 1$ correspond, respectively, to $\frac{1}{i}\{p, \overline{p}\}_{(x,\xi)} > 0$, $\frac{1}{i}\{p, \overline{p}\}_{(x,\xi)} < 0$. Since ([16], [26]) $\frac{1}{i}\{p, \overline{p}\}_{(x,\xi)} > 0$ is a local solvability but non-regularity condition, and conversely for $\frac{1}{i}\{p, \overline{p}\}_{(x,\xi)} < 0$, our choice of test-operator is "justified".

References

[1] AUSLANDER, L. & KOSTANT, B. - Polarization and unitary representations of solvable Lie groups, Inventiones Math. 14 (1971), 255-354 .

[2] BOUTET de MONVEL, L. - Hypoelliptic operators with double characteristics and related pseudo-differential operators, to appear .

[3] BOUTET de MONVEL, L. & TREVES, F. - On a class of pseudo-differential operators with double characteristics, to appear, Inventiones Math .

[4] BOUTET de MONVEL, L. & TREVES, F. - On a class of systems of pseudo-differential operators with double characteristics, to appear .

[5] DUISTERMAAT, J.J. - Oscillatory integrals, Lagrange immersions and unfoldings of singularities, to appear.

[6] DUISTERMAAT, J.J. & HÖRMANDER, L. - Fourier integral operators II, Acta Math. 128 (1972) , 183-269 .

[7] FOLLAND, G.B. - A fundamental solution for a subelliptic operator, Bull. Amer. Math. Soc. 79 (1973) , 373-376 .

[8] FOLLAND, G.B. & KOHN, J.J. - The Neumann problem for the Cauchy-Riemann complex, Ann. of Math. Studies #75, Princeton Univ. Press, Princeton, N.J. , 1972

[9] FOLLAND, G.B. & STEIN, E.M. - Parametrices and estimates for the $\bar{\partial}_b$ comples on strongly pseudoconvex boundaries, Bull. Amer. Math. Soc., to appear.

[10] FOLLAND, G.B. & STEIN, E.M. - Estimates for the $\bar{\partial}_b$ complex and analysis on the Heisenberg group, to appear .

[11] GILIOLI, A. - A class of second-order evolution equations with double characteristics, Thesis, Rutgers University, (1974) .

[12] GILIOLI, A. & TREVES, F. - An example in the solvability theory of linear partial differential equations, to appear, Amer. J. of Math.

[13] GRUSHIN, V.V. - On a class of hypoelliptic operators, Mat. Abornik 83 (125) (1970), 456-473 (Math. USSR Sbornik 12 (1970) 458-476).

[14] GRUSHIN, V.V. - On a class of hypoelliptic pseudo-differential operators degenerate on a sub-manifold, Mat. Sbornik 84 (126) (1971), 111-134 (Math. USSR Sbornik 13 (1971), 155-185).

[15] GUILLEMIN, V.W. & STERNBERG, S. - Subelliptic estimates for complexes, Proc. Nat. Acad. Sci. USA 67 (1970), 271-274 .

[16] HÖRMANDER, L. - Pseudodifferential operators and non-elliptic boundary value problems, Ann. of Math. 83 (1966), 129-209 .

[17] KIRILLOV, A.A. - Unitary representations of nilpotent Lie groups, Uspehi, Mat. Nauk 17 (1962), 57-110 (Russian Math. Surveys 17 (1962), 53-104) .

[18] KOSTANT, B. - Symplectic spinors, to appear .

[19] LERAY, J. - Solutions asymptotiques des équations aux derivées partielles, Seminair sur les equations aux derivées partielles (1972-73), College de France, Paris .

[20] MASLOV, V.P. - Perturbation Theory and Asymptotic Methods, Moskov. Gos. Univ., Moscow (1965) (Russian), translated into French by J. Lascoux and R. Sénéor, Dunod, Paris, 1972 .

[21] MESSIAH, A. - Quantum Mechanics, Vol. I., Wiley

[22] ROCKLAND, C. - A new construction of the Guillemin Poincaré complex, Thesis, Princeton University, (1972) (unpublished) .

[23] ROCKLAND, C. - Poisson complexes and subellipticity, to appear, J. Diff. Geom.

[24] SJÖSTRAND, J. - Parametrices for pseudo-differential operators with multiple characteristics, to appear.

[25] TREVES, F. - Concatenations of second-order evolution equations applied to local solvability and hypoellipticity, Comm. Pure Appl. Math. 26 (1973), 201-250 .

[26] TREVES, F. - Winding numbers and the solvability condition (ψ), to appear.

[27] von Neumann, J. - Die eindeutigkeit der Schrödingerschen operatoren, Math. Ann. 104 (1931), 570-578 .

[28] Yoshida, K. - Functional Analysis, second edition, Springer .

Vol. 309: D. H. Sattinger, Topics in Stability and Bifurcation Theory. VI, 190 pages. 1973. DM 20,–

Vol. 310: B. Iversen, Generic Local Structure of the Morphisms in Commutative Algebra. IV, 108 pages. 1973. DM 18,–

Vol. 311: Conference on Commutative Algebra. Edited by J. W. Brewer and E. A. Rutter. VII, 251 pages. 1973. DM 24,–

Vol. 312: Symposium on Ordinary Differential Equations. Edited by W. A. Harris, Jr. and Y. Sibuya. VIII, 204 pages. 1973. DM 22,–

Vol. 313: K. Jörgens and J. Weidmann, Spectral Properties of Hamiltonian Operators. III, 140 pages. 1973. DM 18,–

Vol. 314: M. Deuring, Lectures on the Theory of Algebraic Functions of One Variable. VI, 151 pages. 1973. DM 18,–

Vol. 315: K. Bichteler, Integration Theory (with Special Attention to Vector Measures). VI, 357 pages. 1973. DM 29,–

Vol. 316: Symposium on Non-Well-Posed Problems and Logarithmic Convexity. Edited by R. J. Knops. V, 176 pages. 1973. DM 20,–

Vol. 317: Séminaire Bourbaki – vol. 1971/72. Exposés 400–417. IV, 361 pages. 1973. DM 29,–

Vol. 318: Recent Advances in Topological Dynamics. Edited by A. Beck. VIII, 285 pages. 1973. DM 27,–

Vol. 319: Conference on Group Theory. Edited by R. W. Gatterdam and K. W. Weston. V, 188 pages. 1973. DM 20,–

Vol. 320: Modular Functions of One Variable I. Edited by W. Kuyk. V, 195 pages. 1973. DM 20,–

Vol. 321: Séminaire de Probabilités VII. Edité par P. A. Meyer. VI, 322 pages. 1973. DM 29,–

Vol. 322: Nonlinear Problems in the Physical Sciences and Biology. Edited by I. Stakgold, D. D. Joseph and D. H. Sattinger. VIII, 357 pages. 1973. DM 29,–

Vol. 323: J. L. Lions, Perturbations Singulières dans les Problèmes aux Limites et en Contrôle Optimal. XII, 645 pages. 1973. DM 46,–

Vol. 324: K. Kreith, Oscillation Theory. VI, 109 pages. 1973. DM 18,–

Vol. 325: C.-C. Chou, La Transformation de Fourier Complexe et L'Equation de Convolution. IX, 137 pages. 1973. DM 18,–

Vol. 326: A. Robert, Elliptic Curves. VIII, 264 pages. 1973. DM 24,–

Vol. 327: E. Matlis, One-Dimensional Cohen-Macaulay Rings. XII, 157 pages. 1973. DM 20,–

Vol. 328: J. R. Büchi and D. Siefkes, The Monadic Second Order Theory of All Countable Ordinals. VI, 217 pages. 1973. DM 22,–

Vol. 329: W. Trebels, Multipliers for (C, α)-Bounded Fourier Expansions in Banach Spaces and Approximation Theory. VII, 103 pages. 1973. DM 18,–

Vol. 330: Proceedings of the Second Japan-USSR Symposium on Probability Theory. Edited by G. Maruyama and Yu. V. Prokhorov. VI, 550 pages. 1973. DM 40,–

Vol. 331: Summer School on Topological Vector Spaces. Edited by L. Waelbroeck. VI, 226 pages. 1973. DM 22,–

Vol. 332: Séminaire Pierre Lelong (Analyse) Année 1971-1972. V, 131 pages. 1973. DM 18,–

Vol. 333: Numerische, insbesondere approximationstheoretische Behandlung von Funktionalgleichungen. Herausgegeben von R. Ansorge und W. Törnig. VI, 296 Seiten. 1973. DM 27,–

Vol. 334: F. Schweiger, The Metrical Theory of Jacobi-Perron Algorithm. V, 111 pages. 1973. DM 18,–

Vol. 335: H. Huck, R. Roitzsch, U. Simon, W. Vortisch, R. Walden, B. Wegner und W. Wendland, Beweismethoden der Differentialgeometrie im Großen. IX, 159 Seiten. 1973. DM 20,–

Vol. 336: L'Analyse Harmonique dans le Domaine Complexe. Edité par E. J. Akutowicz. VIII, 169 pages. 1973. DM 20,–

Vol. 337: Cambridge Summer School in Mathematical Logic. Edited by A. R. D. Mathias and H. Rogers. IX, 660 pages. 1973. DM 46,–

Vol: 338: J. Lindenstrauss and L. Tzafriri, Classical Banach Spaces. IX, 243 pages. 1973. DM 24,–

Vol. 339: G. Kempf, F. Knudsen, D. Mumford and B. Saint-Donat, Toroidal Embeddings I. VIII, 209 pages. 1973. DM 22,–

Vol. 340: Groupes de Monodromie en Géométrie Algébrique. (SGA 7 II). Par P. Deligne et N. Katz. X, 438 pages. 1973. DM 44,–

Vol. 341: Algebraic K-Theory I, Higher K-Theories. Edited by H. Bass. XV, 335 pages. 1973. DM 29,–

Vol. 342: Algebraic K-Theory II, "Classical" Algebraic K-Theory, and Connections with Arithmetic. Edited by H. Bass. XV, 527 pages. 1973. DM 40,–

Vol. 343: Algebraic K-Theory III, Hermitian K-Theory and Geometric Applications. Edited by H. Bass. XV, 572 pages. 1973. DM 40,–

Vol. 344: A. S. Troelstra (Editor), Metamathematical Investigation of Intuitionistic Arithmetic and Analysis. XVII, 485 pages. 1973. DM 38,–

Vol. 345: Proceedings of a Conference on Operator Theory. Edited by P. A. Fillmore. VI, 228 pages. 1973. DM 22,–

Vol. 346: Fučík et al., Spectral Analysis of Nonlinear Operators. II, 287 pages. 1973. DM 26,–

Vol. 347: J. M. Boardman and R. M. Vogt, Homotopy Invariant Algebraic Structures on Topological Spaces. X, 257 pages. 1973. DM 24,–

Vol. 348: A. M. Mathai and R. K. Saxena, Generalized Hypergeometric Functions with Applications in Statistics and Physical Sciences. VII, 314 pages. 1973. DM 26,–

Vol. 349: Modular Functions of One Variable II. Edited by W. Kuyk and P. Deligne. V, 598 pages. 1973. DM 38,–

Vol. 350: Modular Functions of One Variable III. Edited by W. Kuyk and J.-P. Serre. V, 350 pages. 1973. DM 26,–

Vol. 351: H. Tachikawa, Quasi-Frobenius Rings and Generalizations. XI, 172 pages. 1973. DM 20,–

Vol. 352: J. D. Fay, Theta Functions on Riemann Surfaces. V, 137 pages. 1973. DM 18,–

Voi. 353: Proceedings of the Conference. on Orders, Group Rings and Related Topics. Organized by J. S. Hsia, M. L. Madan and T. G. Ralley. X, 224 pages. 1973. DM 22,–

Vol. 354: K. J. Devlin, Aspects of Constructibility. XII, 240 pages. 1973. DM 24,–

Vol. 355: M. Sion, A Theory of Semigroup Valued Measures. V, 140 pages. 1973. DM 18,–

Vol. 356: W. L. J. van der Kallen, Infinitesimally Central-Extensions of Chevalley Groups. VII, 147 pages. 1973. DM 18,–

Vol. 357: W. Borho, P. Gabriel und R. Rentschler, Primideale in Einhüllenden auflösbarer Lie-Algebren. V, 182 Seiten. 1973. DM 20,–

Vol. 358: F. L. Williams, Tensor Products of Principal Series Representations. VI, 132 pages. 1973. DM 18,–

Vol. 359: U. Stammbach, Homology in Group Theory. VIII, 183 pages. 1973. DM 20,–

Vol. 360: W. J. Padgett and R. L. Taylor, Laws of Large Numbers for Normed Linear Spaces and Certain Fréchet Spaces. VI, 111 pages. 1973. DM 18,–

Vol. 361: J. W. Schutz, Foundations of Special Relativity: Kinematic Axioms for Minkowski Space Time. XX, 314 pages. 1973. DM 26,–

Vol. 362: Proceedings of the Conference on Numerical Solution of Ordinary Differential Equations. Edited by D. Bettis. VIII, 490 pages. 1974. DM 34,–

Vol. 363: Conference on the Numerical Solution of Differential Equations. Edited by G. A. Watson. IX, 221 pages. 1974. DM 20,–

Vol. 364: Proceedings on Infinite Dimensional Holomorphy. Edited by T. L. Hayden and T. J. Suffridge. VII, 212 pages. 1974. DM 20,–

Vol. 365: R. P. Gilbert, Constructive Methods for Elliptic Equations. VII, 397 pages. 1974. DM 26,–

Vol. 366: R. Steinberg, Conjugacy Classes in Algebraic Groups (Notes by V. V. Deodhar). VI, 159 pages. 1974. DM 18,–

Vol. 367: K. Langmann und W. Lütkebohmert, Cousinverteilungen und Fortsetzungssätze. VI, 151 Seiten. 1974. DM 16,–

Vol. 368: R. J. Milgram, Unstable Homotopy from the Stable Point of View. V, 109 pages. 1974. DM 16,–

Vol. 369: Victoria Symposium on Nonstandard Analysis. Edited by A. Hurd and P. Loeb. XVIII, 339 pages. 1974. DM 26,–

Vol. 370: B. Mazur and W. Messing, Universal Extensions and One Dimensional Crystalline Cohomology. VII, 134 pages. 1974. DM 16,–

Vol. 371: V. Poenaru, Analyse Différentielle. V, 228 pages. 1974. DM 20,–

Vol. 372: Proceedings of the Second International Conference on the Theory of Groups 1973. Edited by M. F. Newman. VII, 740 pages. 1974. DM 48,–

Vol. 373: A. E. R. Woodcock and T. Poston, A Geometrical Study of the Elementary Catastrophes. V, 257 pages. 1974. DM 22,–

Vol. 374: S. Yamamuro, Differential Calculus in Topological Linear Spaces. IV, 179 pages. 1974. DM 18,–

Vol. 375: Topology Conference 1973. Edited by R. F. Dickman Jr. and P. Fletcher. X, 283 pages. 1974. DM 24,–

Vol. 376: D. B. Osteyee and I. J. Good, Information, Weight of Evidence, the Singularity between Probability Measures and Signal Detection. XI, 156 pages. 1974. DM 16.–

Vol. 377: A. M. Fink, Almost Periodic Differential Equations. VIII, 336 pages. 1974. DM 26,–

Vol. 378: TOPO 72 – General Topology and its Applications. Proceedings 1972. Edited by R. Alò, R. W. Heath and J. Nagata. XIV, 651 pages. 1974. DM 50,–

Vol. 379: A. Badrikian et S. Chevet, Mesures Cylindriques, Espaces de Wiener et Fonctions Aléatoires Gaussiennes. X, 383 pages. 1974. DM 32,–

Vol. 380: M. Petrich, Rings and Semigroups. VIII, 182 pages. 1974. DM 18,–

Vol. 381: Séminaire de Probabilités VIII. Edité par P. A. Meyer. IX, 354 pages. 1974. DM 32,–

Vol. 382: J. H. van Lint, Combinatorial Theory Seminar Eindhoven University of Technology. VI, 131 pages. 1974. DM 18,–

Vol. 383: Séminaire Bourbaki – vol. 1972/73. Exposés 418-435 IV, 334 pages. 1974. DM 30,–

Vol. 384: Functional Analysis and Applications, Proceedings 1972. Edited by L. Nachbin. V, 270 pages. 1974. DM 22,–

Vol. 385: J. Douglas Jr. and T. Dupont, Collocation Methods for Parabolic Equations in a Single Space Variable (Based on C^1-Piecewise-Polynomial Spaces). V, 147 pages. 1974. DM 16,–

Vol. 386: J. Tits, Buildings of Spherical Type and Finite BN-Pairs. IX, 299 pages. 1974. DM 24,–

Vol. 387: C. P. Bruter, Eléments de la Théorie des Matroïdes. V, 138 pages. 1974. DM 18,–

Vol. 388: R. L. Lipsman, Group Representations. X, 166 pages. 1974. DM 20,–

Vol. 389: M.-A. Knus et M. Ojanguren, Théorie de la Descente et Algèbres d' Azumaya. IV, 163 pages. 1974. DM 20,–

Vol. 390: P. A. Meyer, P. Priouret et F. Spitzer, Ecole d'Eté de Probabilités de Saint-Flour III – 1973. Edité par A. Badrikian et P.-L. Hennequin. VIII, 189 pages. 1974. DM 20,–

Vol. 391: J. Gray, Formal Category Theory: Adjointness for 2-Categories. XII, 282 pages. 1974. DM 24,–

Vol. 392: Géométrie Différentielle, Colloque, Santiago de Compostela, Espagne 1972. Edité par E. Vidal. VI, 225 pages. 1974. DM 20,–

Vol. 393: G. Wassermann, Stability of Unfoldings. IX, 164 pages. 1974. DM 20,–

Vol. 394: W. M. Patterson 3rd, Iterative Methods for the Solution of a Linear Operator Equation in Hilbert Space – A Survey. III, 183 pages. 1974. DM 20,–

Vol. 395: Numerische Behandlung nichtlinearer Integrodifferential- und Differentialgleichungen. Tagung 1973. Herausgegeben von R. Ansorge und W. Törnig. VII, 313 Seiten. 1974. DM 28,–

Vol. 396: K. H. Hofmann, M. Mislove and A. Stralka, The Pontryagin Duality of Compact O-Dimensional Semilattices and its Applications. XVI, 122 pages. 1974. DM 18,–

Vol. 397: T. Yamada, The Schur Subgroup of the Brauer Group. V, 159 pages. 1974. DM 18,–

Vol. 398: Théories de l'Information, Actes des Rencontres de Marseille-Luminy, 1973. Edité par J. Kampé de Fériet et C. Picard. XII, 201 pages. 1974. DM 23,–

Vol. 399: Functional Analysis and its Applications, Proceedings 1973. Edited by H. G. Garnir, K. R. Unni and J. H. Williamson. XVII, 569 pages. 1974. DM 44,–

Vol. 400: A Crash Course on Kleinian Groups – San Francisco 1974. Edited by L. Bers and I. Kra. VII, 130 pages. 1974. DM 18,–

Vol. 401: F. Atiyah, Elliptic Operators and Compact Groups. V, 93 pages. 1974. DM 18,–

Vol. 402: M. Waldschmidt, Nombres Transcendants. VIII, 277 pages. 1974. DM 25,–

Vol. 403: Combinatorial Mathematics – Proceedings 1972. Edited by D. A. Holton. VIII, 148 pages. 1974. DM 18,–

Vol. 404: Théorie du Potentiel et Analyse Harmonique. Edité par J. Faraut. V, 245 pages. 1974. DM 25,–

Vol. 405: K. Devlin and H. Johnsbråten, The Souslin Problem. VIII, 132 pages. 1974. DM 18,–

Vol. 406: Graphs and Combinatorics – Proceedings 1973. Edited by R. A. Bari and F. Harary. VIII, 355 pages. 1974. DM 30,–

Vol. 407: P. Berthelot, Cohomologie Cristalline des Schémas de Caracteristique p > o. VIII, 598 pages. 1974. DM 44,–

Vol. 408: J. Wermer, Potential Theory. VIII, 146 pages. 1974. DM 18,–

Vol. 409: Fonctions de Plusieurs Variables Complexes, Séminaire François Norguet 1970–1973. XIII, 612 pages. 1974. DM 47,–

Vol. 410: Séminaire Pierre Lelong (Analyse) Année 1972–1973. VI, 181 pages. 1974. DM 18,–

Vol. 411: Hypergraph Seminar. Ohio State University, 1972. Edited by C. Berge and D. Ray-Chaudhuri. IX, 287 pages. 1974. DM 28,–

Vol. 412: Classification of Algebraic Varieties and Compact Complex Manifolds. Proceedings 1974. Edited by H. Popp. V, 333 pages. 1974. DM 30,–

Vol. 413: M. Bruneau, Variation Totale d'une Fonction. XIV, 332 pages. 1974. DM 30,–

Vol. 414: T. Kambayashi, M. Miyanishi and M. Takeuchi, Unipotent Algebraic Groups. VI, 165 pages. 1974. DM 20,–

Vol. 415: Ordinary and Partial Differential Equations, Proceedings of the Conference held at Dundee, 1974. XVII, 447 pages. 1974. DM 37,–

Vol. 416: M. E. Taylor, Pseudo Differential Operators. IV, 155 pages. 1974. DM 18,–

Vol. 417: H. H. Keller, Differential Calculus in Locally Convex Spaces. XVI, 131 pages. 1974. DM 18,–

Vol. 418: Localization in Group Theory and Homotopy Theory and Related Topics Battelle Seattle 1974 Seminar. Edited by P. J. Hilton. VI, 171 pages. 1974. DM 20,–

Vol. 419: Topics in Analysis – Proceedings 1970. Edited by O. E. Lehto, I. S. Louhivaara, and R. H. Nevanlinna. XIII, 391 pages. 1974. DM 35,–

Vol. 420: Category Seminar. Proceedings, Sydney Category Theory Seminar 1972/73. Edited by G. M. Kelly. VI, 375 pages. 1974. DM 32,–

Vol. 421: V. Poénaru, Groupes Discrets. VI, 216 pages. 1974. DM 23,–

Vol. 422: J.-M. Lemaire, Algèbres Connexes et Homologie des Espaces de Lacets. XIV, 133 pages. 1974. DM 23,–

Vol. 423: S. S. Abhyankar and A. M. Sathaye, Geometric Theory of Algebraic Space Curves. XIV, 302 pages. 1974. DM 28,–

Vol. 424: L. Weiss and J. Wolfowitz, Maximum Probability Estimators and Related Topics. V, 106 pages. 1974. DM 18,–

Vol. 425: P. R. Chernoff and J. E. Marsden, Properties of Infinite Dimensional Hamiltonian Systems. IV, 160 pages. 1974. DM 20,–

Vol. 426: M. L. Silverstein, Symmetric Markov Processes. IX, 287 pages. 1974. DM 28,–

Vol. 427: H. Omori, Infinite Dimensional Lie Transformation Groups. XII, 149 pages. 1974. DM 18,–

Vol. 428: Algebraic and Geometrical Methods in Topology, Proceedings 1973. Edited by L. F. McAuley. XI, 280 pages. 1974. DM 28,–